The R.A.M.S. Library of Alchemy

Volume 30

Five Short Works of Glauber

by

Johann Rudolph Glauber

R.A.M.S. Publishing Company

Five Short Works of Glauber

by

Johann Rudolph Glauber

Translated by

Christopher Packe

Produced by

Restorers of Alchemical Manuscripts Society

R.A.M.S. Publishing Company

R.A.M.S. Publishing Company
117 Rutherford Lane
Stuarts Draft VA 24477

Five Short Works of Glauber
Copyright © 2015 R.A.M.S. Publishing Company

First Edition 2015

ISBN-13 **978-1511697460**
ISBN-10 **1511697466**

Image Processing by Philip N. Wheeler

Printed in the United States of America

Table of Contents

Dedicated to Hans W. Nintzel,

American Alchemist

and

Founder of the

Restorers of Alchemical Manuscripts Society

(R.A.M.S.)

Disclaimer

Liability: The publisher does not warrant or assume any legal liability or responsibility for the accuracy, completeness, or usefulness of any information, apparatus, product, or process disclosed. The publisher makes no representation as to the accuracy or completeness of the contents of this book and specifically disclaims any implied warranty of merchantability or fitness for a particular purpose. No warranty may be created or extended by written sales materials or sales representatives. You should obtain professional consultation where appropriate. The publisher shall not be liable for any loss of profit or other commercial or personal damages, including but not limited to special, incidental, consequential, or other damages.

Introduction

Philip N. Wheeler

The Consolation of Navigators, Tartar from Lees of Wine (also called The Work of Tartar), and Book of Fires were written by Johann Glauber in the 17th Century. I have combined them into this one book because they are each very short.

Also included are Glauber's "A Short Book of Dialogues" and "Novum Lumen Chymicum."

Christopher Packe provided the English translation from the High Dutch. The English edition was published in London in 1689, and from it this work was extracted. Glauber's works are voluminous, consisting of over 2,500 pages of text. This part of the complete works of Glauber was restored and added to the R.A.M.S. (Restorers of Alchemical Manuscripts Society) Library by Hans W. Nintzel in 1982 and 1983.

THE CONSOLATION OF NAVIGATORS

In which is Taught:

How they who Travel by Sea may preserve themselves from Hunger and Thirst, as also from Diseases, which are wont to happen to them in long Voyages.

Written for the help, Comfort, and Solace of all those who make long Voyages for the God of their Country.

THE PREFACE

Reader,

OUR SAVIOR CHRIST hath prescribed to us this Doctrine, that we should behave ourselves towards our Neighbor, as we would that he should do to us; yea, that we should love him as our selves:

This He hath earnestly commended unto us, as the indispensable Will of God, contained in the Law and the Prophets. But although there be few who consider this, and all men seek only their own, yet one or other is still found, who calls this duty to mind, and as far as he can, takes every opportunity or occasion of serving his Neighbor. Although also there may be some who bear a love to others, and desire to assist them with their counsel and help, and yet are destitute of a power of giving them anything; for no man can distribute more than he hath. Moreover, Covetousness year to year, for the public Good, and being affected towards my Neighbor, has made them public. I have also determined, if time shall permit, to publish yet more and better things of this nature. But after that some described Voyages to the EAST and WEST-INDIES, and other far distant places, had happened into my hands, I perceived, not with great admirations what immense dangers occur on the Seas, not only from Robbers and Pirates, but also the adversities of violent Winds, (to pass by many other Calamities in silence) by which the Ship, together with all in it that draw breath, are overwhelmed, and perish: And when I further contemplate those things with an intent mind, nothing seems to me more grievous and intolerable than Hunger and Thirst, or the want of Meat and Drink, which sometimes happens to Ships by adverse Fortune: For when they are overcome by the hand of an Enemy, they are wont indeed to suffer the loss of their Goods, but the Lives of the men are for the most part saved; which loss of Goods they may again easily repair by Merchandizing; and although they should be adjudged to death, yet their pain would quickly have an end. But if through an ill fate of necessity one be destitute of Bread and Water, and be forced to tear another in pieces to eat, that is of all the highest misfortune and death itself were more desirable, than to sustain extremities and miseries of this sort. And although the want of Food doth not so frequently happen, yet the want of Water often falls out, whence Seamen are grievously tormented with Thirst, to whom, as in a most urgent Calamity, Mercy and Christian Compassion ought to be administered; but, Who can carry them succor afar off, and in a tempestuous Sea? And seeing that the Prosperity of Maritime Countries (where for the most part there is no Wine, no Fruits, no Mines, as in the upper GERMANY, and other places) consisting only in Navigation, it were very well worth the while that this should be promoted to the utmost. I have therefore diligently considered the matter with myself, and have found that a Remedy may be applied to this Evil, viz. the want of Meat and Drink, by providing in time an APPARATUS, or certain matter to be carried in Ships, which is of far greater efficacy in mitigating HUNGER and THIRST, than common Bread and Water; yea, is able to prevent and expel that Disease which is familiar to Seamen, to wit, the SCURVY, which is often a great hindrance to Sailing: So that this matter may be carried together with the usual Provisions, as a Preservative, and in case of necessity, as if the Ship be overmuch harassed with Storms, or be hindered by other infelicities, or detained longer in the Voyage than expected, it may be taken and used. It were indeed to be wished, that these materials might never be wanted in any Ships, but as the old Proverb hath it, A SUPERFLUOUS CAUTION NEVER DOTH HURT; therefore it is always better to have a thing in readiness, which we do not use, than to want it when its use is necessary. But what those materials are, of which I speak, and how they are to be used when need is, I shall communicate to my Neighbor, for the public Good, not at all doubting, but that this invention will prove highly profitable to many who use the Seas.

The Reader now understands what hath moved me to write this Treatise, viz, the love of my Neighbor, and that I have not com posed it only for the sake of some few, but that I have emitted it to the public to the end, that those highly profitable Invent ions might afford help and comfort to all Mankind, and especially to all that use the Seas, and such who are infirm in their health:

Nor do I doubt but that this my sincerity of mind will be a great preservative, comfort, and relief to those who pass the Seas, when they are in danger, by the benefit of which, they may escape various Chances and many Misfortunes, or beware of them for the future:

For by this means which I shall here produce against Hunger and Thirst, or other Diseases which are wont to afflict Sailors, it will be found, that what I ascribe to them may be fully deduced to the desired effect. And therefore many Thousands of men may thence receive Fruit and Profit, as long as the World shall endure; so that all who travel the Ocean, and are afflicted with any Disease, ought to rejoice, and give Thanks to God.

Now, if any one should yet doubt of the success of this matter, (which nevertheless is showed from so clear and irrefrangible Fundamentals) I leave him to his freedom, whether he will put the matter into examination, and certify himself of the Truth, before he give credit to it: Which also may be proved in a small quantity, and not only on the Seas in Voyages, but on the Land also in the House, among both the Sound and the Sick. Therefore let no man vilify what he doth not understand, but let him commit the thing to trial, and set its reason and quality, before he condemned it, or pass an immature Judgment, lest his curiosity or imaginary Wisdom from a vain instinct, deceive him, or confound him with shame, When the proof shall show it to be better than he could persuade himself.

But that in some places I have used obscure words, and have not proposed all things so clearly, as that every man will be able to understand them, let no man wonder at this for I have certain reasons for so doing; for no man will have any prejudice or injury thereby, seeing that nevertheless there are such points of this sort elucidated, which will remain after me safe and sound; Therefore whatsoever the Reader shall here find written, let him esteem it worthy to be received and embraced, as the Gift of God; the which, If I find to be gratefully accepted, more (God willing) shall follow. Also to all those who shall have need, I offer the Medicaments aforesaid, profitable against Hunger and Thirst, and all Sea faring Diseases, a good quantity of which I will cause to be prepared, that every man may use them that will, and thence satisfy his desire.

Wherefore it is the interest of any to whom I have committed the Preparation of this Medicine, to betake himself to it, and when need shall be, he will not be unwilling to use it. I have not sought myself in these things, being content to serve my Neighbor out of Christian Charity. And although this my good Will shall not be accepted by foolish and ungrateful men, nevertheless God shall have the praise, who hath committed to us a mutual participation of love and good will, which also in his own time will grant the benefit of this to be derived and redound to my Children after me, by some pious Souls, of which I make no doubt; and therefore readily acquiesce in my present condition.

The Consolation of Navigators, etc.

Now to come to the Work itself, we will point out the Remedies, whose use is so necessarily required in Navigation, and which is able to preserve us not Only from the pressures of Hunger and Thirst, but also from the injury of Diseases; and they are no other than Corn and Water concentrated, or reduced into a more compact and narrow compass, the one for the extinguishing of Hunger, the other of Thirst; and how they are both to be concentrated; and administered in case of necessity, I will exactly describe and teach. And:

Part I. Of the Concentration of Corn or Grain.

As for this, it is sufficiently explained in the FIRST PART OF THE PROSPERITY OF GERMANY; so that it might be here passed over; nevertheless I will make this short repetition.

Make a Malt of Wheat, Barley, Oats, or other Grain, as is usual for the brewing of Beer, (See the full Description, PROSPERITY OF GERMANY, p. I, cap. 2.) and extract all the strength with Water, as if Beer were to be made thereof: Afterwards boil this Liquor away gently in broad and shallow Vessels or Coppers, to the consistency of Honey: The Dregs or Grains serve for Food for Cattle, but the inspissated liquor or juice may be commodiously carried by Sea, and at pleasure may be made into

Beer with a mixture of Hops and Water. And because for the most part eight Tons of Grain afford one Ton of the inspissated Juice, every Ton of Grain makes a Ton and an half, yea, two Tons of Beer, for every Ton of Liquor makes at the least eight, ten, twelve, or more Tons of Beer, according as you will have it stronger or smaller. Therefore it is easier and cheaper to carry in a ship one Ton of this Juice, then ten or twelve Hogsheads of Beer, which easily corrupts and grows sour; but on the contrary, this Liquor being kept from the Air, retains its goodness; and this is of singular advantage, seeing that good fresh Beer may be made of this juice. To this also belongs another great Commodity, viz, that if this Juice be mixed instead of Water, with Wheat-flower, and baked, it makes an efficacious Bread, which affords far more nourishment to sailors, than their common Bread, and contains in itself so great power, that it can refresh and cheer the Sick: For which reason our Ancestors did not in vain use to temper fine Flower with clarified Honey instead of Water, and make Bread thereof, which they called Cakes of Life, because they were a great support to the Humane Body and, as it were excited the Life itself: But in our days all things being subservient to AVARICE, you may see those sweet Cakes, made up with common, impure, and unclarified Honey, which cannot gene rate much good Blood, or juices; but

this our Bread will prove itself to be of a greater
sweetness, nobility, and efficacy, in as much as the
elicited juice of Grain, affords a far better
nourishment than Honey. But if any desire to proceed
farther with it, this concentrated Liquor of Corn
will yet afford a greater utility, viz, if being
inspissated or brought to a thick consistency, it be
mingled with fine Flower of the best Malt, and Bread
made thereof, which after It is baked, is to be cut
into pieces, and again put into the Oven, till it be
dry and hard, and then put up into Chests or
Hogsheads, to preserve it from the Air, and so
carried to Sea; for then this Bread, when necessity
requires, may be infused in warm Water with a few
Hops, and excited to the separation of its faeces,
and it will acquire the substance of Beer: But that
which doth not pass into the Liquor, nor become
Beer, may be heated in a Kettle, and some Butter put
to it, which will very much comfort the languishing
stomachs of Seamen, as well, or better than the
eating of Bread Softened in Beer. But in regard that
bitter Potions are not agreeable to all Palates, it
may also be made into Beer without Hops, by mixing
the bread with good water only, and afterwards
boiling it for the evaporating the more phlegmatic
part, which will have a pleasant taste, But this
Biscuit also, or twice baked Bread, may be ground
small in a Mill, and put up close in Casks, and

preserved on shipboard. Afterwards, when need shall require, you may temper it with warm water, and in an open Hogs-head; which yet must have a bottom, suffer it to settle, and clear itself; so the flower ascends upwards, and the water attracts the sweet liquor, and hence becomes excellent and wholesome Beer, which if it be drawn out at the bottom of the Hogshead runs as clear, as if it had been made some months; for there is now a separation made of the pure part of the Bread, from the impure or gross, and when that which is fine is drawn off, the gross part may also be drawn out of the Vessel, which being boiled with butter, affords a singular pleasantness to the taste; yea, also is of a greater salubrity, and better digestion in the stomach, than Pease, Beans, or French Barley boiled: So that here is nothing lost, Bread or Meal of this sort affording good Beer, and also a wholesome Food to eat. And thus on shipboard one may at any time of the year not only have good fresh Beer, but also thence may be made good Vinegar. And this may suffice to have briefly taught the manner of mitigating Hunger and Thirst by concentrated Corn.

I shall now treat of the Cure of Diseases, to which Seamen are liable, and which often bring death.

Now if you regard the Nature of Man, you shall generally find that he uses no measure or mediocrity in eating and drinking, but rather puts down so much till his belly will hold no more. Which Vice one man obtrudes upon another, under the show of kindness and good will, from an old custom and enormous abuse, although in this one offends more than another. Therefore while the Appetite is more liberally indulged, than admits of a good digestion, the Liver assumes a Chyle which is scarce half concocted: Whence also a gross blood is generated; and so in process of time the Viscera are filled and obstructed with crass and viscous humors; hence they cannot perform their office as they ought, but rather many and divers Diseases do arise, according as the heap of crudities encompass the flesh covering the Joints, obstructing the Veins and Nerves, and deprave all the MEDIUMS of nourishment; therefore when the Evil cometh to that pass, and grows prevalent, the whole body is sensible of it, but chiefly in that place where it fixed its seat:

And hence necessarily one part suffers by the hurt of another, till at length all the powers go to decay, and the whole body languishes, and cannot help itself, and unless succored by Art, dies.

For these causes Physick was invented, that those Diseases arising from intemperate eating and drinking, might be met in the way, the perverse Enemy or primogenial Vice of the body be removed, the viscous and phlegmatic SALARRA of the depraved humors dissolved and opened, and the oppressed Members expurged, that so the body may again recover its former healthful estate; the which is wont to be done by divers means, and in various manners, according as the Physician understands the Disease to be, so he also affords his help, viz, taking it away by Evacuations upwards or downwards, by Sweat or Urine, or by other means, as the nature of the Disease requires. But by which way soever the recrements of the malignity be dissipated and expelled, and the inwards parts be freed from them, it is well and the Physician hath performed his Office, also merits Thanks and a due Reward. Therefore whosoever well understands the nature, rise, and birth of diseases, and also possesses a good Medicine, or can obtain one, is fitly qualified for a Physician; but he who knows not the disease, nor is also furnished with fit and efficacious Medicines, nor knows how to obtain them, grievously errs by trying Experiments so long, till the Evil more and more increasing, the Sick at length expires; the which is so well known that it needs no proof; so that many who have contracted diseases from immoderate eating and drinking, are afraid to commit themselves to an unskillful Physician, but choose rather to cure themselves by abstinence and fasting, which way is indeed safe, although it be tedious and full of delay.

Others again use vulgar and trifling Medicines, and yet recover their health, tho' late, thinking that this happened by the use of their Medicine, when nevertheless the length of time, and the small portions of their meat and drink, during their illness, whence the superfluous pravity of humors hath gradually wasted, hath effected this; which also sometimes happens to those who take no Physick; but by how much the greater the pravity of the humors is, by so much the longer it will be before Nature will be able without help to overcome and expel them. The which, if it take 4, 5, or 6 weeks to accomplish, a Physician (by the benefit of good Medicines) might effect it in 2 or 3 days. So great is the difference between a Cure which Nature performs in a long time, and that which is quickly done by Art.

But here it may be objected, that all diseases take not their beginning so much from an evil and superfluous humidity, as from manifold other causes, one being derived from this, another from another cause. To this I answer, that all those affects of the Body which exist in the skin, and do not proceed from external accidents, as wounds, bruises, falls, & etc. whence the bruising of the body, and also Death, at length happens, do draw their original from the intemperance of eating and drinking, which administer occasion, and furnish matter to noxious

24

humors, which hence occupy and infect divers Members, for one part affects another, till the whole body abounds, and is imbued with a multitude of viscous humors. When the Stomach is over-gorged, and its tone is spoiled, it contracts cold crudities, and falls into a dangerous estate. How then should it rightly digest the Food? And, what good can it transmit to the Liver? And seeing the Liver receives nothing but what is evil and depraved, what can that procreate of good, and diffuse throughout the whole body? Therefore as I have said, one Member must suffer by another, till the error become common. Whilst a Tree or an Herb in the ground remains temperate, that it be neither too wet nor too dry, its root draws from the earth virtue, and imparted it to the trunk, and the trunk to the branches, leaves, flowers, and fruit, and is able to produce fruit conducible to health, for many years. But if the root be ill placed, the Salt being ill disposed, whatsoever the stock there finds, it associates to itself, and also communicates the same, and no better to the branches: And if it stand too wet, it brings forth an insalubrious Fruit, which by reason of too much moisture, falls off before it is ripe, in whose stead fungous protuberances arising from putrefaction, spring up, and such Plants do not endure long.

Again, if the Root stands too dry, it cannot
thence have juice sufficient to nourish the Tree,
and bring forth Fruit, but will by degrees wither
away and die. So also it is with Men, and their
Diseases; for according as their Bodies are treated,
they are healthful and fruitful, or else disease.
Now seeing that I have proved Diseases from
redundancy or humidity, or from dryness, it will be
easier to provide preventive Remedies against them;
or if any error or delay should happen in this, it
may be amended in the curative part, by which the
Evil may be abolished. These are required to be of
that nature and property, that they attract the
superfluous and pernicious humors from all the
Viscera and principal Internal Members of the whole
Body, into the Ventricle, conciliate a new
concoction or digestion, separate the pure from the
impure, adjoin the one to the Liver, and eject the
other by Siege: And so the body is not only freed
from depraved burdensome humors, but also recovers
its strength, and is cured of all Infirmities. Which
manner of Curing, at this day, the more the pity, is
known but to few; seeing that it is not to be
effected with common Herbs, but somewhat better is
required, than what the Philosophy of old Women
prescribed: Nevertheless there is a great efficacy
in certain Simples; as may be seen in HELLEBORE or
HELLEBORASTER, by whose benefit the Ancients

prolonged their lives, by the daily use of a certain dose of it. Moreover, Tobacco where it cometh to maturity, effects things to be admired, being rightly prepared and administered; yea, even that which (being crude) is taken in Pipes, cools and refreshes the body, and also in some measure relieves against Hunger and Thirst; which common experience daily witnesses. But whence this Virtue happens. Tobacco takers neither know nor care, but acquiesce, in that they either receive pleasure or profit by it: Therefore, If Tobacco, or any other vulgar Plant can perform so many and so great things, being yet crude, without any preparation, What would not an Extract or concentrated Essence of all the Vegetables do, being rightly made? Which nevertheless ought to be of that nature, that it may not only dispel every heap of malignant humors, but also strengthen the inward parts, and preserve from all things which may happen in the generating of a Disease.

Such a Medicine is that which I here present to all those who continually use the Seas, and besides the Scurvy, undergo many other Diseases, by which they may not only efficaciously protect themselves against the assaults of Distempers, and hinder their progress when already began, but also under the present want of Meat and Drink, may make sound and infirm body: But how, and from what Ingredients this

Euporist may be made and prepared, needs not to be manifested to everyone. This is a great Gift of God, which ought to be honored, and not prostituted to the Unworthy. Let it suffice at this time, that such a Medicament may be obtained at a small price; I will not suffer it to be buried with me, but will leave it to others, who may keep it, and sell it to those who desire it at a reasonable rate.

This is given in form of an Electuary, and may be taken for a Preservative, daily or every second, third or fourth day, as occasion requires, in the quantity of half or a whole Pease, upon an empty stomach, fasting two hours after it, if need be, but if not, Food may be taken presently after it, although it is better to abstain some hours: But when a Disease hath already invaded a man, whether it be FEVER, SCURVY, HEADACHE, CATARRH, or any other Disease infesting Seamen, let the Patient forthwith swallow down of this Electuary the magnitude of an ordinary Pease (for it hath no unpleasant taste) and sweat if he can; if not, let him keep however a very moderate Diet all the day, and in Summer time, as much as he can, avoid the intense heat, and in Winter, the extremity of Cold; the next day let him take the quantity of a Pease or two, and so let him increase or diminish the dose, according to the state of the Disease. These things being well observed, all Sick nesses will give place.

If a man carries with him but half an ounce of this Medicine to Sea, in a long Voyage, he will possess a PANACEA, both for the preventing and curing all Diseases incident to his body.

This is of very great use and profit especially for Masters of Ships, to whom I also highly recommend it as such, and not to them only, but also to all the Inhabitants of the Earth, to whom it promises no less success. If a Medicine can be invented, (besides that universal one of the Philosophers) accommodated to all diseases, truly this is one, and will perform all, or even more than I have attributed to my CATHOLIC in the SECOND PART OF MY PHARMACOPEIA SPAGYRICA, that is a Powder, but this, of which I now discourse, an Electuary prepared of certain good Ingredients and Sugar. I affirm again, That there is no disease, whether internal or external, for the curing of which this Medicine doth not suffice; for it doth not only resist the FEVER, SCURVY, and POX, but even the GOUT and LEPROSY itself, provided they be not too much fixed and radicated; yea, although they be very inveterate, and cannot be thoroughly rooted out by this Medicine, nevertheless it gives relief in them, and renders them more tolerable, and hinders the accession of the daily fits, and keeps it under. Certainly where this Medicine can yield no help, Nature must be wholly ruined and tired out. What I have here written, is the Truth, which I am able to prove by experience; however it is free for everyone to believe or not believe me, as he pleases: Christian Charity to my Neighbor hath compelled me, if it be well taken, it is well, if not, I have done my part, and discharged my Conscience. Nor will I insist any farther on the praise of this PANACEA, I have said enough to those who believe, or can understand me.

And thus I have set before the whole World, that Medicament which can preserve from, and cure the most grievous Diseases accompanying Ships and Voyages, and also relieve in the extremity of Hunger.

This being done, there yet remains another Remedy whereby the Body may be relieved, and refreshed in extremity of Thirst, when drink is wanting; and this is done by Water concentrated, which is of so great virtue, that it gives such a refrigeration to a thirsty Palate, that in the whole World there is not the like. Seeing therefore that in long Voyages Water is often wanting, and putrefies and stinks, it will be very necessary here to declare and show, how Thirst may be restrained and rendered tolerable in time of necessity. Then also how common Water may be preserved from corruption; moreover, that one Ton may effect as much for the quenching of Thirst, as otherwise 2 or

3 Tons of common Water can. Which is not only my saying, but the Truth itself; as I will here prove by manifest testimonies: And this is to be done as I have said, by Water concentrated and coagulated. But what that is, I will first show, and then by what means it is to be prepared and used.

To explain the genuine Property of concentrated Water to many, may seem unnecessary, in regard that Seamen are not skilled in such discourses, nor careful about them; and that they rather belong to Philosophers, and the Contemplators of Nature, that they may render the knowledge of them familiar to themselves: But this Treatise without doubt will come into the hands of wise and experienced men, as well as into those who use the Seas, and therefore perhaps I shall not lose my labor, if I somewhat more exactly describe the nature of that water.

It is sufficiently evident to all those who have saluted but the Thresholds of Nature, that the Omnipotent God, in the first Creation of the World, hath separated the Elements from the rude CHAOS, as also the Elements from one another, plating the Earth in the bottom or foundation of the Center, and over that the Water, over the Water the Air, and over the Air the Fire; so that every of them holds its own receptacle and seat, from whence, without the will of God, it cannot recede. Nevertheless, we find that one always participates of the other, and none is found without another, although one is predominant and visible; but the rest exist in it invisibly, and by the skill of an Artist, may visibly be deduced from it. So from the Earth we elicit Air, Water, and Fire; from the visible Water, natural Earth, Air, and Fire; again, from the Air, Earth, Water and Fire; and from the Fire, Air, Water, and Earth. Hence the Elements are continually circulated from one species into another, which indeed we do not observe or perceive, and no Element can want an Element, but draws its life and nutriment from another. The Fire cannot burn without Air; the Fire operates upon the Water, and of it makes Air; the Water rests in the Earth, and moistens the same; the Fire impregnates the Air; the Air insinuates the Seed into the Earth; the Earth nourishes and cherishes the Seed unto perfection, and afterwards brings to light what it hath conceived, and carried in its belly.

These things are only hinted at by the way, and not without cause.

But to return to our concentrated water, that we may explain its nature, know, that water is the PRINCIPIUM or beginning of all the Elements; which thing is sufficiently manifest, and may be seen daily, especially in the SUBTERRANEAN Mines of Metals, whose inward bowels are so penetrated, that there is not only water, but we also see it to be quickly transmuted into various forms of Mineral Bodies, the which without is familiar; and the more limpid or clear the water is, the brighter Stones and purer Minerals it generates: Of which you may read more in that Treatise, (THE SECOND PART OF THE MINERAL WORK) where I have described the generation of Minerals; also that Flints, and Sand, in the Earth, Rivers, and Seas, receive their increase from water, is sufficiently manifest; for all sand was primitively water only, and hath passed out of water into hard sand or stones, and into which it may again be changed. But this is not to be used as a Remedy against Thirst, because it is now too hard, and difficultly returns to its first matter, viz. Water, except by the benefit of another subject, which is a MEDIUM between Sand, Flints, Crystal, and common Water, viz. Salt, which hath an agreement with the nature both of common water and that of stones, and may be easily changed into either of them; as you shall hereafter hear. But this concentrated Water, which I here propose for the vanquishing of Thirst, and refreshing of the body, is salt prepared and concentrated, either of the waters of Mountains, or of the Sea, which are of the same goodness: Of which afterwards is prepared by Art a water so excellent, that for repressing of Thirst and refreshing a languid body, there is no Remedy in the world may be compared to it.

Here someone ignorant of natural things may object, and say, How can it be that Salt should repel Thirst, seeing that when it is eaten, it excites and increases the same? To this I answer, That common salt, as it is first extracted or boiled, can in no wise restrain the Appetite of drinking, but rather provoke it, except its pravity be first corrected and depurated; for every salt seems to have laid aside its nature and property, so long as it contains earthy and gross faeces, which nevertheless may be removed from it by solution, filtration, and coagulation; whence also it is bitter, astringent, and hard; but this harshness and terrestreity being removed, the salt remains no longer hard, but is reduced into a soft and fluid Liquor, not sweet, but pleasantly acid, tasting like an Apple or the stones of Grapes. And this purification must be made by the force of Fire, viz, when the salt in distilling is sublimed in earthen Vessels, by strong Flames; and then the purest and most noble part only ascends, and the more gross and earthly remains in the bottom, and is of no use; but the purer part is of great power and virtues, not only against Thirst, but is also very necessary for many other things, as shall hereafter be demonstrated. That otherwise in impure common salt a great Virtue is latent, let no man doubt, seeing that it may be daily observed: Wherefore that

excellent Philosopher PLATO writes, THAT IN SALT
THERE IS SOMEWHAT DIVINE; and according -to the
opinion of other great and wise men, GOD hath openly
manifested Himself in Fire and Salt, and the ancient
Philosophers were hence taught to prepare their
UNIVERSAL MEDICINE; which Art of preparing they have
called ALCHEMY, or THE MELTING OF SALT; but I speak
not this by experience, in regard that I never yet
durst attempt so great a Secret, being contented
with smaller things; but this I know, and have
divers times done, viz, that by a certain Artifice,
I have precipitated from common Salt some corporeal
Gold, although without profit, and only to see the
possibility, which without those precipitations
remain spiritual, and at length by them obtains a
corporeal substance.

Indeed the greatest Mystery of the whole World
consists in Fire and Salt, and with those two (after
God) nothing may be compared; for the Flame raises
Light out of Darkness, which otherwise none but God
can do. It is also fit to restore life to the dead,
no otherwise than the warm Sun or Oven revives
Worms, Frogs, Serpents, Flies, and other Insects,
which have been killed by the Cold; of which I have
made mention in the second part of my PHARM. SPAGYR,
as well as also in MIRACULUM MUNDI, and its
EXPLICATION. Many men use Fire, yet know not what
they have or use: So also it happens with salt, all

men use it out of custom, but observe not what they taste. Salt is a thing of great moment, by it we give strength to resist Corruption, both to the living and the dead.

And so it is, that if men had not salt, or at the least those Plants and Fruits, in which it dwells, they would putrefy while yet living. Which would also happen to Cattle, if they did not eat Grass, which contains in itself its own salt, and suffices them in feeding; nevertheless if something better were given them with it, it would be very profitable for them; for if to horned Beasts, as also Hogs, some saltiness were given with their meat and drink, they would thereby acquire much strength and fatness.

Seeing therefore that Salt, whilst it is yet gross and rude, and defiled with many impurities, performs such things in living Creatures, and preserves Flesh and Fish, as also all Herbs and Grass, long from Corruption: What could not such a salt do, which by the help of Art hath received a greater purity than Nature hath bestowed upon it? For the Fire contains a power of correcting and purifying salt, and converting it into a finer and better sub stance. The common salts which we daily use, do indeed give a' relish to all Edibles, and preserve them from putrefaction; yet nevertheless they dry, astringe, and bind the belly; also excite

Thirst; but salt being rectified and depurated, communicates a sweet and grateful taste, also more powerfully resists Corruption than the other, neither doth it bind the belly, but keeps it soluble, expels Urine, nor doth it cause Thirst, but rather takes it away both from the sick and sound: So great a difference is there between the common and corrected and depurated salt, which Rustics little understand. The nature and property of common salt is known to everyone, as far as concerns the Kitchen, but the rectified is known to few: And the Spirit of Salt, which is in use in Apothecaries shops, and which they also sell, being for the most part not rightly prepared, effects little, and therefore not frequently used: But if it were made and rectified as it should be, it would not only have a pleasanter taste, but also a better operation. So many things bear Names which are not agreeable to the thing, and therefore cannot perform what is expected from them.

The Ancients have ascribed great Virtues to Spirit of Salt in Medicine, which it indeed hath, if after its distillation it be well rectified and dephlegmed; but if it be administered so crude as it ascends the first time, it may perhaps do more harm than good, Inasmuch as a great deal of terrestreity comes over with it, which in the rectifying remains in the bottom of the Glass, as an insipid, bitter, or astringent salt. And although this is well enough known to some, nevertheless they abhor to rectify their spirit, because the spirit in rectification loses a great part of its weight, by separating its superfluities.

I taught, many years since, how to prepare the spirit of salt in quantity, whose description is also exactly set down in the FIRST AND SECOND PART OF MY FURNACES. It also demonstrates its utility in Metallics; as manifestly appears in the FIRST PART OF MY MINERAL WORK, where the manner of extracting Gold from Stones, by its help, is taught. Although the spirit for that work should be prepared after another manner, than that which is to be used at Sea, or on shore, for the curing Diseases or repressing of Thirst; wherefore I'll first teach the way of making this, and then also show its Use.

As for the manner of anatomizing Salt by Fire and Art, and trans-muting it into a pure and sweet spirit, it is various, for one uses this, another way, as I also have taught divers ways, In the FIRST AND SECOND PART OF MY FURNACES; but here. I will appoint another, which is the best and most commodious for the uses treated of in this Book, and is thus to be performed.

Get some strong Retorts made of good Earth (but Glass is better) fill them with the following mixture of Salt and Earth, then according to Art,

distill thence a spirit gratefully acid, which duly
rectify, and it will be prepared for Use. The salt
is to be thus prepared for distillation: Fill a
great Crucible with Sea salt, which cover with an
earthen Cover, and by degrees put Fire about it of
flaming Coals; increasing it, until the salt be all
over red-hot; this being done, presently remove it
from the fire, and let it cool; by this means, if
any greasy sordes, or other combustible impurities,
adhere to the salt, they will be consumed and
vanish. Dissolve the salt in common water, filter or
strain it through a fine close Linen Cloth, that the
faeces may be separated; the clear Liquor evaporate
in an earthen glazed Vessel, till the Cuticula
appear, then make little balls of clay and sand,
about the bigness of Walnuts or Pigeons Eggs; dry
them, and heat them red-hot in an earthen Vessel,
and so the fat spirit of the Earth will be burnt
away; then imbibe those balls with your strong salt
water, by letting them remain in it for some hours,
that they may drink in as much of it as they can,
then take them out, and dry them: With these fill
your Retorts, and administer Fire by degrees, and at
length urge with the strongest flames, so long, till
they emit no more fume, for then all the spirit is
come forth: The Receiver must be of Glass, and very
large, in whose bottom must also be a little water,
to attract and condense the spirit. The Distillation

being finished, the spirit is to be rectified by itself in a low Cucurbit in sand, and first the water or phlegm will come over, (which you may keep for another distillation) then the pure and sweet spirit will rise limpid and clear, and an unpleasant salt will remain in the bottom, which is to be cast away, and the spirit to be kept for the uses to be showed hereafter.

But if you desire to have the spirit yet more pure and efficacious, you may rectify it again upon calcined or powdered Flints, which will retain in the bottom whatsoever is yet gross or impure, and the spirit will be rendered very pure and subtle; for the Flint at the first was Water and Salt, and hence both contain in themselves so straight a communion. This may be seen among the Glass makers, who reduce Sand or Flints into clearness by melting, that thence they may make pure Glass; for Salt in the Fire dissolves Sand, Flint, Crystals, and together with them receives the Essence of Glass; yea, and these very species and matters, by the help of an ALKALI Salt, are resolved into a diaphanous Glass, which being put into common Water, melt like Salt; of which I have discoursed in the THIRD PART OF MY FURNACES. When we distill the aforesaid rectified Spirit of Salt again from powdered Flints, from the innate love which it hath to the Flints, it carries some part of them upwards with itself, and is made better for use in all Dis eases, especially in the DROPSY, STONE, and GOUT it is an excellent Medicine. The Flints remaining in the bottom, yield a water which dissolves the hardest Crystals, and is acid like the spirit of Salt, and nothing else but part of it coagulated with them, and may be administered in all cases as the Spirit of Salt.

Let no man wonder that I ascribe more to this Spirit rectified with Flints, than to any other common spirit of salt; for few know what Virtues are latent in flints or sand. All Birds and four footed Animals use sand. Many men administer the powder of Flints and Crystal, for the corroborating of weak stomachs. Hens devour Sand and small Pebbles, that they may thence have matter for their Eggshells, and being debarred of it, lay Eggs without shells. Wherefore I affirm, that this spirit of Salt, abstracted from flints, to be better in many Diseases than the common spirit and especially the Crystals, or spirit of salt coagulated, of which we have above made mention.

Let this suffice concerning the Preparation of spirit of salt. Now we will also teach, how it is to be administered either at Sea or on Shore, for the repelling of Thirst, and the Cure of Diseases:

To what other purposes the spirit of salt serves, besides Medicine, I have showed elsewhere; here my intention is only to signify of what great

41

use and profit it may be in Voyages to Sea, which was the only occasion of writing this Treatise.

Of the Use of SPIRIT OF SALT in Ships, against Thirst, and also the Scurvy.

In the first place, this spirit is profitable to the body, when a small quantity of it is put into the Hogsheads of Water, whereof Seamen and Soldiers drink, to which it gives a grateful acidity, like that of Wine, and is far more effectual in quenching of Thirst, and in less quantity than other Water, and besides preserves the Water from corruption and stinking, which otherwise in long Voyages it is wont to suffer, contracting a muddiness, blackness, and breeding Worms, which generate the SCURVY; for the Water containing the spirit of salt, cannot corrupt; which spirit also preserves other things, both living and dead, from putrefaction. And because the water with the acid spirit of salt extinguishes Thirst more than any other, it necessarily follows, that so great a quantity of water need not be carried for a Voyage, as otherwise is usual. Moreover, seeing that the spirit of salt resists all putrefaction, it suffers not the SCURVY to take root, for it refreshes and exhilarates a man with great vigor, corroborates the stomach, and all the members of the body, it consumes piteous matter in the Reins and Bladder, expels Urine and Stone, keeps the Belly soluble, promotes digestion; it suffers no Disease to grow from corruption of the Blood, to which otherwise Sea-faring men are subject: Moreover, it exerts its salutiferous Virtues, if a little of it be mingled with Wine and Beer, in Ships, which will long preserve it in its goodness and sincerity, causing it the better to repress Thirst, and therefore to go the farther in use. Also Beef, Mutton, and Veal may by it be preserved a long time, viz, if they be put into stone pots, and powdered or condited with some salt, dissolved in the spirit of salt. Nevertheless it behooves that the Flesh be without many Bones, and that all the legs, shins, or shanks be cut off. Moreover, a very small quantity of the spirit being added to the water in which Fish is to be boiled, makes the Fish eat firm and close, and of a much solider relish than when sprinkled with Vinegar. So also Raisins being put into water impregnated with spirit of salt, will grow soft, and swell, and will recover their strength, and sweetness, as if they were fresh plucked from the Vine, which will afford a great refreshment to those who frequent the Seas.

In a word, this spirit may commodiously be used, not only in Water, Wine, or Beer, but also in all sorts of Food in Ships; for it gives a very pleasant taste to all meat and drink, and correct eth them so, that they become more wholesome. It hath also another very profitable use, viz. If in hot weather but one drop, or so small a part of the coagulated spirit, as the magnitude of an Hempseed, be mixed with sugar and held upon the Tongue; for it so cools the mouth, and allays Thirst, that there will be no need of bad Water or Beer, by which the body is more violated than corroborated, through drinking often and much of them. Nor is this the least of its virtues, that when in a long Voyage there is a want of water, a little more of the spirit of salt may be put into it than will serve for a grateful acidity; for so it will quench Thirst for a longer time, and in a less quantity than otherwise a greater quantity of water can do. Yea, although the Ship should be cast away, and the men forced to betake themselves to their Boat, or broken pieces of the Ship, as it often happens, if they have but an ounce of the spirit of salt in a Glass, or half so much of it coagulated, in defect of Drinks, they may sustain their lives and health therewith so long, till it please God to afford them a means of getting to shore; nor will they be forced to drink their own Urine, or Sea-water, which is pernicious.

These and the like Virtues the spirit of salt show in Ships, which for brevity sake, are not all declared.

I have also meditated how the SEA-WATER may be made sweet and drinkable by precipitation, which would be very desirable in want of water: And so far I have ATTAINED, that I can precipitate a good quantity of the salt from the water, which also becomes much sweeter, but yet not so fully, but some saltiness remains in it. Although in case of necessity this water may be used in Ships for the boiling of Pease and Beans, as also for the fermenting of CONCENTRATED CORN. What I now know concerning this precipitation, I will make manifest for the good of my Neighbor, and show a good beginning how Sea-water may be made fit to drink.

There is a certain kind of Mineral called MARY'S ICE; it is not the MUSCOVIAN Glass, as perhaps some may think, but another thing; when it is heated red—hot in the fire, and so cast into Sea-water, it is presently reduced to a tender and white powder; which being done, the Water will be agitated and moved hither and thither, then the Powder attracts to itself part of the Salt from the Water, and settles to the bottom of the Vessel, and the Water becomes clear, which may be poured out, and if this labor be once or twice repeated, the Water, by this precipitation, will indeed be made sweeter, but not altogether potable. Also the salt or solution of SATURN dismisses much of the salt from Sea-water, yet by that it is not rendered drinkable. The best way which I now know, is this which follows, but it is somewhat costly:

Nevertheless when necessity urges, sweet and good water profits a man more than a Bond of an hundred pieces of Gold. Hence this Art deserves to be honored, although it be costly; for a man ready to die with Thirst would give all he hath for a draught of Water to save his Life: Wherefore somewhat of this precipitation may be committed to Ships, for their preservation, which may be used in great necessity, but if no such be, it may be brought home again, and kept for another Voyage, seeing that it will in no wise decay, but be as good after an hundred years, as the first day of its preparation,

But what this is, there is no need openly to divulge, but he who desires it, may come to me and have it: But for the information of those who are skillful in CHEMISTRY, I will intimate this:

That the precipitation of Salt from Water, is necessarily to be performed by a singular sand, which doth not only drive salt down-wards, but also all Phlegm, Sordes, and Impurities; so that although the water should be like to a Fen or Dunghill in filth and stink, by the benefit of this precipitation, in a few hours, it should become like clear Fountain water, because the evil odor and taste would also be precipitated.

The same may be done in all Waters, how nasty and muddy soever, and not only in Waters, but also in all potable Liquors, as Wine and Beer, although the Wine were red, seeing that the precipitation casts to the bottom all muddiness, color, and stink. And hence not only those Liquors, which are changed red by corruption, or other accident, but also Crystals, which are red by nature, may be reduced to a perspicuous clarity; which is a thing very profitable for Vintners; for otherwise the Sea-waters become sweet when they are carried far thought common sand, which imbibes their salt; for those two have a mutual communion and communication, seeing that both are generated of Water; hence also by the fire, they are resolved into a dry Water, or pellucid Glass. Whence any man knowing in the nature of things, may apprehend, that this precipitation of Sea-water is built upon a foundation agreeable to Nature.

My ALEXIPHARMAC, or Mineral-Electuary, before mentioned, as also the Spirit and Crystals of Salt, may be administered, as well at Land as on Sea, against the SCURVY, FEVERS, and other Diseases.

And although in the SECOND PART OF MY PHARMACOP. I have already described a certain Mineral Remedy, under the Title of a PANACEA, and commended it against all Diseases, by which admirable Cures are everywhere done, yet I prefer my

Mineral Electuary before that, because it is more accommodated to use, seeing that when need requires, it may be taken out of a Box with a Knife only; neither is there need of any Vehicle, as Wine, Beef, or others, which are necessary to be used with the Panacea; neither is there need of weighing it, but may be proportioned by the Eye, according to the age and strength of the sick. Those who are of full age and strength, may take as much of the Electuary upon a fasting stomach as will equal the magnitude of a Pease, and fast some hours after it, if time and opportunity permit; also let the Patient avoid the cold Air, and the heat of the Sun in hot seasons. It is to be taken twice, four times, or more often, in a week, according to the necessity of the sick, to which it must also be proportioned by increasing or diminishing the Dose; for if one grain should not work, the Patient may take two, ascending to 3 or 4, if need be. When a small Dose is administered, it works insensibly, if a little more, then it works by stool, and sometimes by vomit, when it finds disease matter in the stomach; if it finds depraved salt humors, it casts them out by sweat, spitting, or urine, and thoroughly purges the body from every evil. In brief, I say, it greatly purifies the Blood, opens all obstructions of the internal principal members, as the LIVER, SPLEEN, and LUNGS, beyond all PHLEGMAGOGUES: It hinders the growth of all APOSTEMATIONS both within and without the body; it consumes FLUXES and CATARRHS, which fall upon the Eyes, Ears, and Teeth, it exterminates the venereal POX in a short time; it also cures the DROPSY, LEPROSY, GOUT, AND FALLING-SICKNESS, both in old and young; expels the STONE both of the KIDNEYS and BLADDER; cures all sorts of FEVERS; and lastly, it heals all inward and outward WOUNDS and ULCERS, being taken inwardly, and a due Diet observed.

This is a safe and approved Medicine in the cure of all curable diseases, nor doth it fail to help in inveterate affects, viz. when a disease hath been long growing, and is now firmly rooted and fixed, that it cannot be totally expelled, as the GOUT and STONE, yet then this Medicine mitigates the pain, and makes the Fits slower and more tolerable, and suffers not the DISEASE to increase, but to be more and more abated and diminished. It cures all SCABS and RINGWORMS, or TEETERS, proceeding from an impure Blood and Liver, only by taking it inwardly, without any outward unctions.

What shall I say more? This Medicine cannot be sufficiently praised, seeing that it operates universally against all the diseases both of Men and Beasts: But I must forbear, for if I should show all its Virtues, this Treatise would swell to too great a bulk.

In MARITIME Towns, and other places where the SCURVY is regnant, a better Medicine than this cannot be found, for it removes all pravity of Humors, from which such diseases proceed; it preserves from the PLAGUE, and other Contagions, and happily cures their Infections; it strengthens the Memory, Heart, and Brain; preserves from the PALSY and its Concomitants: In time it restores those who have their Members contracted, and their Nerves repressed and immovable, after the being anointed with Mercury, for the cure of the POX. He that will take it every week or once in 3 or 4 weeks as a preservative, shall not have the TOOTH-ACHE, nor shall his Ears be troubled with ringing or other noises, nor shall his Eyes be troubled with sharp Rheuins, but by the benefit of this, those continual inveterate Corroders of the Body, CATARRH, FISTULA, CANCER, and other almost incurable symptoms, are thoroughly cured.

Hence both Rich and Poor, and those especially who go long Voyages to Sea, and want Physicians, ought to esteem this ALEXIPHARMIC, and provide themselves with it against a time of necessity. If any man carries with him half an ounce of it, it will be sufficient for preservation and curation for more than a year, in many diseases. If a Ship going to the EAST or WEST INDIES, carry with them half a pound of it, they may save the Lives of many men;

one of which is of ten times more value than the price of the Medicine, by help of which the lives and health of many Hundreds of Men may be pre served, and being sick, may be reduced to their former health.

Now it is farther necessary to describe the Virtues which the Spirit of Salt shows on Shore; for that which is profitable to the sick on shipboard, is not harmful for those who dwell at Land: For this spirit may be used in all Houses in their usual drink, as Beer or Wine, especially in these moist Regions, where the Scurvy commonly reigns; so that there is scarcely an House to be found, in which there is not one at the least infected with this Disease; for the drink is hereby made clear, so that it well dilates the Urinary passages, moreover hinders the growing of the stone in the Kidneys and Bladder; it also gives a pleasant taste to the Wine, removes its superfluous Tartar and precipitates it to the bottom, so that SPANISH and FRENCH Wine acquire a clearness like to RHENISH wine. In Summer time, after Liquors have once wrought, it suffers them not to fret, grow mucous, or work again, but keeps them from many Accidents. FRENCH Wines as sold in many places, have neither odor nor taste, but good Wine is endowed both with a Vinous relish, and a grateful odor; and when these are wanting, the name agrees not with the thing. But the Spirit of

Salt, with the Essence of Wine, conciliates to
FRENCH wine the smell, taste, and color of RHENISH
wine. Hence also being so meliorated, it will, like
RHENISH, keep many years, which otherwise will
hardly hold good two years. Moreover, by the spirit
of salt that quality may be taken away from Honey,
which renders its taste somewhat ungrateful, and its
hidden Impurities may be precipitated, so that an
excellent wholesome drink may be made of it. Which
is a famous Art to be used in those places where
there is no wine; for when Honey is well depurated,
and the unpleasant taste taken from it, it becomes
as a comforting Balsam to Man; as that old Solder
had experienced, whom ALEXANDER asked what he had
used to prolong his Life to so great an Age, he
answered, That inwardly he had used Honey, and
outwardly Oil. And it is evident to many, that there
is a great virtue in Honey, but by reason of the
unpleasant taste, it is loathed; which yet is
removed by the spirit of salt, so that a drink
prepared of such pure Honey, is as wholesome as the
best wine: which hath also this benefit, that every
Housekeeper may have this drink at his Table, at any
time of the year. This honeyed Nectar also holds
good many years (may be preserved for many years -
pnw), after the manner of the best RHENISH wine, and
is made for a small charge, and will be a great
comfort to those who are not able to go to the price

of wine, Moreover, every one may make his own drink as strong and as sweet as he pleases; which is the principal head of the Art, to take away the ungrateful taste from the Honey, which being removed by the spirit of salt, the drink will become clear, to which you may add a little FRENCH or RHENISH wine for taste sake. You may thence also make good Beer, which will hold its goodness 10 or 20 years, or more, viz. if instead of common water you take, the water of Hops, to dissolve the purified Honey, to which (if you please) you may give divers tastes, by putting into it divers Herbs, as is wont to be done in wine and beer. But this you are especially to regard, That for this business you do not use unrectified spirit of salt, for such would spoil the taste, and corrupt the drink, but always take that which is well rectified, which I have sufficiently taught the manner of doing of both in this Treatise, and divers other places of my Writings.

N. B. There is no need to put the spirit into a Vessel of Wine or Beer, but it suffices to keep it in a small glass, and as often as you have occasion to put some drops of it into a Bottle of Wine or Beer, and so to drink of it. Every one therefore may make his drink as he will, by putting in more or less of the spirit, according to his Palate; which doth not only serve to make the wine and beer clear and diuretic, but may also be used in Fountain-water, in hot weather, for the allaying of wine: For if some drops of it be put into a bottle of water, it gives it a grateful acidity, not much unlike to the natural Mineral acid waters, and in hot weather allays the over great heat of the Blood, and quenches Thirst, so that there is no need of burdening the stomach with much wine or beer, Also all those drinks, as wine, beer, and water, which are mingled with the well-rectified spirit of salt, are far more wholesome, than other ordinary drinks. If a little pure Steel be dissolved in spirit of Salt, and then put to fair water, it will have almost the same taste with the SPAW-WATER, and if some quantity of it be drank, it loosens the belly, and evacuates black Excrements, whence a man grows lively and vigorous, as if he had drank of the SPAW-WATERS.

And this benefit may accrue by it to all Travelers, who carrying with them a little spirit of salt in a glass, may at all times, and in all places, correct and amend their beer or wine in their Tuns (a large cask for liquids, especially wine -pnw), and give it a better relish: But because this spirit is more liable to be lost in a liquid form by any accident befalling the glass, therefore being coagulated into a dry form, it may be more commodiously carried about in a Paper or Box, of which the magnitude of a small Pease drank in one cup of water, quenches Thirst better than a great quantity of water alone. This spirit of salt will be serviceable to men of all sorts, but to none more than to Seamen or Soldiers, in war-like expeditions, of which sometimes more die through want of water, than by the sword, who also for the most part contract their Diseases from the filths in Ships, which might easily be prevented. What will it profit a Captain of a Man of War, or the Commander of a Merchant-ship, to have many men for his defense and service, if they are disabled by a Disease? Were it not then better to use all diligence to preserve their health, and to restore it when lost? Which may be done with small Labor and Cost. A General brings an Army some time into the Field, of 10, 20, or 30 Thousand Men, and seeing that those observe no due order in eating and drinking, and being destitute of Medicines, if the PLAGUE, FEVER, or BLOODY-FLUX comes among them, they drop off one after another like Flies in Winter; when notwithstanding they might be succored and helped in those Diseases by my Electuary, (This Electuary I think to be the PANACEA ANTIMONII made up with Sugar). Salt of Salt, or Spirit of Salt, inspissated or thickened with Sugar. The spirit of salt hath always been highly esteemed, both by ancient & modern Physicians, for its great virtues which it shows both inwardly & outwardly; besides which, it is of great use in other Arts, as

I have showed in the FIRST AND SECOND PART OF MY FURNACES. And that the Reader may see that I am not singular in what I have ascribed to the spirit of salt, I will here set down the Experience of that no less learned than famous Physician CONRADUS KUNRATH, of the spirit of salt, as he hath expressed it in his MEDULLA DESTILLATORIA, printed at HAMBURGH, ANNO 1638; in these words, Part I, page 59.

The Great and Effectual Operations which are attributed to the Spirit or Oil of SALT.

The Spirit or Oil of Salt is not of so acrimonious a taste, as Salt commonly is: Nor hath it so great an acidity as the Spirit of Vitriol, but it tastes almost like the juice of wild Apples, with somewhat of sweetness intermixed. It discusses, dissolves, consumes, and dries, yet it doth not heat too much, but hath a temperate nature, mitigating and comforting the natural heat, which it also increases, and dispels all things contrary to Nature, preserves the state of sound humors unhurt, especially profitable for those who are phlegmatic, whose viscous SABURRA of humidities it consumes, takes away CATARRHS which fall from the Head, and preserves from all Diseases which arise from a noxious Pituity, Mucor, and Fluxes. Yea, those who use labor and study, shall perceive their Strength to be restored by it, and as it were renewed.

Experience, the Mistress of things, witnesses that it is a present Remedy against the FALLING-SICKNESS, being administered in good AQUA-VITAE: Against the APOPLEXY, PALSY, or loss of SPEECH; as also the trembling and beating of the Heart, and all dejections of Mind; as also in the time of the PLAGUE, or in any infectious Air, it may be administered with profit, by mixing half an ounce of the Spirit or Oil of Salt with two ounces of the Juice of Violets, and as much of the Electuary of Juniper-berries, in a Stone or Glass Mortar, and the Patient taking the quantity of a Filbert in the morning upon an empty stomach.

This Compound Medicine being given to Women with Child, facilitates the Birth, and is highly profitable against various symptoms of Women in Child-bed, without any danger. If any desire to preserve their native heat, let them often use this ALEXIPHARMIC.

Whoever is burdened with superfluous moisture, let him take this Spirit or Oil of Salt daily, in Wine or some other Vehicle.

In Coldness, and continual Fluxes, which stop up the Breast, it conduces much, and takes away inveterate Coughs and grievous ASTHMA'S, which are stirred up by those Fluxes. It dissolves, consumes, and dissipates depraved humors, collected and indurated in the stomach; and although of itself it communicates little nourishment, yet it excites the Stomach to a good Appetite, and disposes the Meat to a good digestion and passage through the body.

It helps in the Tartarous and indurated Obstructions of the Liver and Spleen, which it opens and relieves, and hence removes and takes away the JAUNDICE, DROPSY, HYPOCHONDRIAC MELANCHOLY, and pains and effects of the RIBS and VISCERA, and also whatsoever arises from Wind and Flatulency, and other symptoms proceeding from the said causes, it especially absumeth the ANASARCA, or Water under the skin, and thoroughly removes watery Tumors in the Genital Members or Legs, which for the most part happen to those who are Hydropical, Phthisical, or labor under a Distemper of the Liver. It also takes away the great Thirst in those, which otherwise in this kind of diseases doth not lightly afflict, so that the Patient shall have no desire of drinking, but may remain some days without. The same is also a Remedy against all putrid FEVERS.

In tormenting Pains of the Belly, and COLIC, which are excited by the viscosity of humors, or in temperature of flatulency, or force of cold, and a dense and tough Phlegm driven into the Intestines; this spirit resolves and consumes, and so opens the stopped passages of the Bowels, and loosens the Belly, that it eases the Iliac pains, whether it be taken at the mouth, or administered Clysterwise. In the LIENTERY, DYSENTERY, or HEMORRHAGE, it is to be given in Clysters; in like manner in NEPHRITIC Diseases, and the dolorous Stone in the Bladder, which it dispels and exterminates.

In RUPTURES, as the HERNIA and ENTEROCELE, let some drops be given daily in good Wine; let the Tumor of the SCROTUM be also often anointed with this Spirit or Oil, and cherished by a fit Truss, or some other bandage, and in a few days the sick will recover his former health.

It kills all WORMS in the Body, casts them out and prohibits their breeding again.

It is most effectual as a preservative against the contagion of the PLAGUE; and also is very conducible to those who are already infected. It also wonderfully profits such as have eaten Poisonous Mushrooms or Opium, as also those who are hurt by Scorpions, Vipers, Spiders, and the like venomous Insects, it is accommodated both to inward and outward use, because it consumes virulent humidities. For the stinging of Wasps and Hornets, it is to be applied topically.

For fat Women, and those who are troubled with impurities of the Matrix, from a superfluous phlegmatic humor, it is very profit able; for by the benefit of it, every redundancy and incommodity of Phlegm is purified, consumed, and dried up, so that the seed may more easily rest in the Womb, and fertility be greatly promoted.

In protuberant Excrescencies of the Eyes, FILMS, CATARACTS, BLISTERS, PUSTULES, OR SPOTS, DIMNESS or DARKNESS OF SIGHT, it is to be used in a good Colliery or other commodious Vehicle. In BLOWS, BRUISES, or FALLS, when the Blood is congealed, (which we call black and blue) in the Eyes or Face, let a bit of a Sponge or Lint be wetted with the Spirit or Oil of Salt, and applied to the place affected; or else dissolve in the same a little red Myrrh, and with the Dissolution mix a little Honey, and anoint with it. It drives away noise and pain of the Ears, also when the Ears send forth an ulcerous and purulent matter; the same Medicine may be administered with a happy success. And in these Symptoms it is to be mixed with Wine-Vinegar, and dropped into the Ears, and rubbed upon the diseased parts.

In the THRUSH, and other sore mouths of sucking Children, also in the glandular or kernelly Swellings of the Neck, foulness of the Tongue, swelled and rotten Gums, SCURVY in the Mouth, TOOTH ACHE, superfluous humors and filths adhering to the Teeth and Gums, this Spirit or Oil of Salt is an excellent Remedy; as also in preserving from those Evils, it is to be conjoined with clarified Honey, and the Mouth, or parts affected in it, to be

anointed therewith. Against the evil Affects and
Ulcers of Women's Breasts, some linen rags are to be
humected with the Spirit or Oil of Salt, and they
will be quickly healed. The moist ITCH, TETTERS,
RINGWORMS, and other contagious Affects of the Skin,
are happily cured by both the inward and outward use
of this Spirit. ULCERS and fungous TUMORS in the
Venereal Disease, or others, if we would bring them
to a separation, the Oil of Salt will effect it,
being applied either alone, or mixed with other
convenient Remedies. For the healing of venomous
Ulcers, which pierce the Bone itself, as also all
virulent Apostems, the Spirit or Oil of Salt is to
be mixed with White-wine, and Barley-flour, into the
consistency of a Plaster, which is to be applied to
the Sore.

Also in cancerous, eating, and spreading
Ulcers, it is very profitable, being mixed with the
juice of Rhue, and aptly applied.

That I may summarily express its use both
inwardly and outwardly, it resists all fluid and
corroding Vices, and Lays the foundation of their
Cure. In luxated, shortened, or contracted members
and joints, it affords present help, being used
alone, or joined with fit fomentations or Epithems.

In the taking away of WARTS, this Spirit is to
be mixed with the juice of Marigolds, and applied to
them: It also takes away Corns, if after they are

softened with warm water, they be well cut and anointed with this Oil. In the ERYSIPELAS, ST. ANTHONY'S FIRE, this being mixed with Elder-Vinegar, and applied, is of great virtue.

The Hands of Feet being corrupted or chapped by Cold are recovered by the benefit of this Medicine. This Spirit or Oil greatly conduces to help the weariness and weakness of the Feet or other Members restoring to them their strength and vigor, if they be well bathed therewith before the fire,

Moreover, in the dolorous GOUT it is a famous Remedy, and a profitable ANODYNE for allaying the pain, if besides its internal use it be also applied topically: For to this our Medicine are to be joined Oil of Turpentine, Oil of Wax, Oil of Chamomile, or also Oil of Cowslips, with which the parts affected are to be well anointed. But where the members are contracted by FLUXIONS and CATARRHS, fit Fomentations are also to be used, and besides our Oil or Spirit, the dulcified Oil of Vitriol, and pure Oil of Turpentine, are to be mixed together, and the parts to be therewith anointed before the fire. And hence the Nerves and Joints are so comforted, that they will not so easily admit such Fluxions again. Also if there be TOPHES or NODES in the Joints, they will be discussed beyond belief, being duly anointed with this Oil, mixed with the dulcified Oil of Vitriol.

It is also used with great profit both inwardly and outwardly, against the CRAMP and CONVULSION OF THE NERVES by Cold.

In external Wounds and Symptoms there often happens a putrid odor, and also proud Flesh sometimes starts up, and causes great pain. These Protuberances being anointed with this Oil or Spirit are presently dissolved and consumed, and also preserved from future corruptions.

In brief, this is a most excellent Medicine, overcoming many Diseases. Moreover, the Reader is to know, that this Spirit or Oil of Salt, (besides those Virtues which I have declared) if it be diligently prepared, and rightly prescribed, may be used for the preparing of many excellent and precious things in the Chemical Art; for it dissolves Gold, Gems, and other Stones, Pearls, Corals, & etc. So that they are reduced into excellent Medicines, in a liquid or potable form, highly useful for Mankind: But how those Processes are to be instituted, I shall here pretermit. But he that hath drawn a good foundation of working from that in formation which my MEDULLA DESTILLATORIA hath faithfully propounded, let him weigh the matter with an accurate mind, and put his hand to the Work, there is nothing in it so tedious, but he will easily conceive; moreover, let him associate himself with honest, sincere, and skillful Artists, and take their counsel, then he will find out many Secrets of Nature, and singular Arcana's, and will see that his care and diligence hath not been in vain: Then let him use that Experience to the Glory of God, and the help of his needy Neighbor.

The Virtues or most efficacious Operations of the SPIRIT or OIL OF SALT, in which Gold is rightly dissolved, according to Art; or when it shall be made an AURUM POTABILE.

Philosophers and Physicians, endowed with the great Exercise and Knowledge of Things, have attributed to the Spirit or Oil of Salt, in which Gold is rightly dissolved, or the AURUM POTABILE made with it, very great operations in the body of Man, inasmuch as in all Diseases and Infirmities, of what name soever, it gives present help, and in all dejections of the vital spirit, although they tend to the fatal period, it gives such relief, that life and vigor may be yet somewhat farther protracted, if two, three, or four drops be administered as occasion shall serve in a good AQUA VITAE or Cordial Water, In like manner, if three drops be administered once a week in generous Wine, or AQUA VITAE, or other fit Vehicle, it renovates a man, makes him youthful, changes gray Hairs, produces new Nails, and Skin, preserves from various and diverse symptoms of Diseases, and preserves the body in such a state, even to the prefixed hour of the Divine appointment.

These are the very words of that famous Philosopher and Chemical Physician CONRADUS KUNRATH, in his MEDULLA DESTILLATORIA; as the Reader may there see.

Now any may easily conjecture, that although this learned man had found out many things, yet that he knew not all, but what was known to him, that he divulged. But that as yet more might be done by that Spirit or Oil of Salt, than what he had set down, easily appears from that which he shows at the end, concerning the oil or Liquor of Gold, prepared by the Spirit of Salt, which words are Truth itself, and much more may be done by the benefit of that solution. He hath written what Experience hath taught him, the rest he hath less to the study and searches of others.

Seeing therefore that I also (not to speak boastingly) have often handled Furnaces and Coals, and among others, have found this Liquor of Gold or AURUM POTABILE, made with Oil of Salt, to be of great virtue, and knowing its great use, power, and virtue by experience, I will describe it in an open stile, lest so noble a Medicine should be buried. Such a Liquor of Gold as is here mentioned, Is compounded of the purest Sol, and the most highly rectified and again concentrated Spirit of Salt: The

Sol is to be first melted, and thoroughly purged
with Antimony, then to be dis solved in AQUA REGIA,
and precipitated with MERCURY, to be edulcorated and
brought into a subtle Calx, which must be heated red
hot (to free it from the MERCURY) and then dissolved
in strong and well-prepared Oil of Salt; being
dissolved, abstract part of the spirit of salt from
it, and a very yellow Liquor of Gold will remain at
the bottom, which yet is not, fit for use alone,
seeing that the Oil of Salt contains too much
Acrimony; therefore a drop of it is to be mixed,
with a spoonful of Beer, Wine, or warm broth, before
it be administered to weaken the spirit of salt; but
if any desire to have it sweeter, instead of Wine,
Beer, or Broth, it may be mixed with melted Sugar,
or syrup of Roses. The Dose for a man of ripe age,
is two or three drops, which if he shall perceive
not to operate sufficiently, he may increase to
three or four drops, so long, till he shall find an
evident operation, which appearing, let him increase
the Dose no longer, but rather diminish it a drop;
and when the Oil of Gold rightly performs its
operations, these signs will appear: In the first
use, a certain loathing or nauseousness of the
stomach, will be perceived, when the Oil of Gold
finds there a vitious impurity, and endeavoring to
expel it, drives it downwards, and ejects it by
stool.

Part 2

The Excrements are as black as a Coal, and
during the use of the Medicine, the sick makes a
discharge by stools at least twice, sometimes three
or four times, without any impediment or sense of
necessity, as is wont to be in the working of
Purges. The Urine will also be thick and turbid,
because the Medicine dissolves the Tartar and Phlegm
in the Reins and Bladder, and by degrees expels it.
N. B. That by the black Excrements, it is manifest,
that Gold may be radically dissolved in the stomach
of a Man, which some think to be impossible. The
humane stomach hath a greater power in the
destruction of things than the strongest fire, as
may be here seen by the Gold: Yea, all things which
are eaten or drank, in the space of 24 hours, it
thoroughly dissolves and transmutes into a new form
of Excrements.

If the stomach of a man can effect such things,
why not Art also? Yea, hence it is expressly proved,
that the colors of Gold, at length, when it is
radically dissolved and destroyed, do appear, and
may be known, seeing that in Colors Blackness hath
as it were the first and chief place, and contains
all others hidden and concentrated under it.

N. B. That these black Excrements should not be
cast away, but the radically dissolved Gold ought to

be separated from them, with which perhaps some great thing may be effected.

I some time since administered this Oil of Gold, for eight or ten days successively, to a certain Infant, for the freeing his body from Mercury, which had been imprudently given him for the Worms; I ordered the Excrements to be saved, for some Experiment, which nevertheless, because they stood long and bred Worms, I could not use, but commanded them to be put to the Roots of a young Vine, which had not yet born Grapes, being but of two years growth, which produced a small Grape with large stones, which had golden spots like Stars, admirable to behold. This Example is worthy of a profound consideration. It also seems to me, that the Urine of those who continually use the aforesaid golden Liquor should be auriferous, although it appears not in the color. The thing is out of doubt, seeing that men, in the use of the said Medicine, attract only a certain hidden virtue, from the Gold, and again send the rest forth of the body, that that Gold is better than other common

Gold: The Benignity of the Divine Being hath disposed and ordained everything in the World for the best.

Seeing that in the Stomach of Man or Beast the Food is destroyed and putrefied, Nature taking a little from it only for nourishment sake, casts out the rest by siege, which yet is not of so abject a condition, as to want its virtues? For if these Excrements be mixed with any Earth, moistened with Rain-water, and exposed to the open Air, there will thence spring forth divers Herbs, without the addition of any seed; but if the seed of any Herb be adjoined, then is also brought forth fruit of the same substance and quality; so that these Excrements may degenerate, and be converted into whatsoever Fruits we will: From which Fountain the multiplication of Herbs and Metals may perhaps take its original. Therefore it is necessary, that Putrefaction should go before Multiplication; which our Savior himself told his Disciples, saying, EXCEPT A GRAIN OF WHEAT SHALL FALL INTO THE EARTH, AND DIE, IT REMAINS ALONE: BUT IF IT DIE, IT BRINGS FORTH MUCH FRUIT. The Axiom of Philosophers is, WHERE NATURE ENDS, AND LEAVES THE WORK IMPERFECT, THERE ART OUGHT TO BEGIN. But the manner of proceeding farther they have involved in silence. Nature from the beginning hath sublimed a Mineral ENS, or first matter of Metals, as much as it could, and brought it to the royal seat, or highest perfection:

Art hath destroyed the most perfect body of Gold by corrosives, and being destroyed, hath again dissolved it in the body of Man:

But if any know the manner of proceeding farther with it, he may easily reduce that Essence of the dissolved Gold into a better, and multiply it: But I would not have this taken for an Oracle, seeing these are only my Cogitations.

The Use of this Medicine is to be continued till the body is freed from all ill symptoms; nor are any other Remedies to be inter mixed with it, that its operation may not be hindered: But before this Medicine be administered, a Dose or two of my PANACEA ought to be given, (of which the SECOND PART of my PHARMACOP. treats) for then its effects are to be admired.

This Oil of Gold, or AURUM POTABILE, is of much greater force in all those Diseases in which the simple spirit of salt is conducible, because the Oil of salt hath double the strength of the spirit, and also disuses its virtues much better, by reason of the Gold which is anatomized in it; the which, because it is of a hot and dry property, agrees very well with the Oil of salt, whose nature is hot and moist: and hence it effectually resists all Corruption which may arise in the humane body.

Besides, the Virtues which the spirit of salt, and especially the Oil of Gold prepared with it, manifest both within and without the body; It is an egregious preservative against Drunkenness, NOTE-Spirit of Salt, a Remedy against Drunkenness.- which is the root of many Diseases, and the gate by which many evils enter; For strong Wine or stale Beer being drank, suffer nothing in the stomach, because it cannot restrain their subtle spirit, which presently flies up into the Head, and disorders and confounds all the senses. But if in the drinking, a little spirit of salt be mixed with the Wine, it opens the Orifice of the stomach, and suffers not the spirits to fly upward, but restrains and binds them, also mitigates and represses that unnatural Thirst, which otherwise the Wine causes in excessive drinking: Nor doth the Wine, which is mixed with spirit of salt, so easily hurt the Liver; for as the spirit of wine heats and inflames the Liver, so the spirit of salt corrects and cools it. Add that the spirit of salt permits not the liquor to lie lurking in the stomach, but presently casts it out by Urine: and the sooner the superfluous Wine is cast out of the body, the less hurt it can do. But this is to be understood of the spirit of salt well rectified, and subtly prepared.

In short, I might sooner want Paper to write, than matter to express what good the spirit of salt coagulated can do, being taken in drink; I have said enough for the present; in my VEGETABLE WORK I will more fully discourse of this matter, in which I now acquiesce. Whatsoever I have here written of a Medicine against all Sea-faring Diseases, and of the concentration of Corn and Water, against Hunger and Thirst; as also of the most efficacious use of spirit of Salt, against Drunkenness, and of the melioration of Wine, Beer, Water, and other Drinks, is built upon the pure foundation of Truth, which Experience itself will sufficiently testify; with which I put an End to this Discourse.

TARTAR FROM LEES OF WINE

A TRUE AND PERFECT DESCRIPTION
OF EXTRACTING GOOD TARTAR FROM THE LEES OF WINE,
And that after a Plentiful Manner.

To the most Reverend PRINCE JOHN PHILIP, Lord Arch-Bishop of Moguntina, HIGH CHANCELLOR OF THE EMPIRE, AND ELECTOR, BISHOP OF HERBIPOLENSIS, DUKE OF FRANCONIA, & etc., my most Honored Lord.

I Presume, you remember (most Reverend and Noble Lord Arch-Bishop and Elector) that about two Years since, I craved Your Lordship's License, to Extract TARTAR from the Lees of Wine; which you Highness was pleased to grant me. From which Work, seeing I have been hither to hindered by the multiplicity of other business, and the great Waste and Loss which is daily sustained in the Electorate of MOGUNTINA, and Bishoprick of HERBIPOLENSIS, proceeding from the Ignorance of the Lees, daily troubles me; I have determined to dedicate this gainful EXTRACTION OF TARTAR FROM THE LEES OF WINE, to your Electoral Reverence, under whose Patronage I may be safe, and for this Reason especially, Because none of the Princes of GERMANY enjoy a greater Vintage than your Reverend Highness; to whom not only a good part of the Tract of the RHINE, but almost the whole Jurisdiction of MAENE, together with FRANCONIA, by Divine Providence, belongs; being very fruitful in Wines; where also much Lees are cast away, through Ignorance OF THE WAY of extracting their TARTAR, for the Good of the Country, from which a good Spirit of Wine might first be drawn by Distillation; the TARTAR contained in the thick Lees, and remaining in the bottom of the Still, is by all Men cast away, as unprofitable, a few excepted, who with great Labor dry them, and of them make clavellated Ashes: Which Waste or Loss of the Lees, seeing that it makes every year an incredible Loss of TARTAR, I have thought fit to expose the Knowledge of this Mystery to the Eyes of all Men, for the Good of my Country, by which some being instructed, may set about the EXTRACTION OF TARTAR with great Gain, not doubting, but many, when they shall see this Labor to be profitable to others, undertaking the same, will bring great Profit to their impoverished Country.

Therefore I trust this Little Work will be acceptable to your Reverend Highness. And I pray the Giver of all Good Things, to furnish both your Soul and Body with His Gracious Benefits, who study to be

Your Highness's most

Humble Servant,

J. Rud. Glauber.

A True and Perfect Description of Extracting good TARTAR from the LEES of WINE, etc.

First of all, we must know what Dregs are, or what is their Essence and Nature, how they are resolved into Parts, and the Good separated from the Un-useful, because without the knowledge of the thing we can't give a right judgment of it, but must err:

Therefore it is necessary that we know what we have in our hands, that we may the safer handle it. For this cause I shall show the Ignorant what are properly Dregs, and which way the better part may be extracted, that hereafter so great a Good may not by ignorant sluggishness be laid aside, but converted to the profit of all.

All thick matter, whether it be of Wine, Beer, Vinegar, or the like, when it has stood a little, sends the earthy, heavy, and more thick part of it, to the bottom, which we call Dregs or Lees, upon which the clearer part flows, to be separated from the impurity, as in the making of common drinks may be seen.

There was no use of these Dregs heretofore, except the making Brandy with them, and the rest to be thrown out of doors, in which was much Virtue, which Men did not think of: But that all things are not unprofitable which the Vulgar pronounce so, and that there may be great Virtue in ordinary Matter, I shall demonstrate by this Excrement of Wine.

When the Juice out of the Winepress is put into Hogsheads, that there working, the dregs falling to the bottom, the clean Wine may come out, the superfluous Salt of new Wine, while it is squeezed out of the Grapes, with the same labor is separated, part sticking to the sides of the Vessel, which we call TARTAR; but the greater part of this Salt or Tartar, implied in the turbid dregs like Sand, sinks to the bottom. Besides, this is the property of Salts, that by a hot humor they make a thin water, the humor growing cold, not being able to keep all the dissolved Salt that is compelled to separate, which excluded the Vehicle, seeks a new place. If you put into the Solutions Sticks or Strings, that Crystalline Salt, in a cubical or angular figure agreeable to its nature, will stick to them; or otherwise it cleaves to the sides of the Vessel.

'Tis above and beyond all Controversy, that the Grape above all Vegetables has much Salt, not sensible, unless it is fermented with a minute heat, which working by Nature, endeavors a separation, while the purer and more liquid part retains so much Salt as the Wine has need of, the thicker Salt being extorbated, part of which incrustates the sides of the Vessel, part of much thickness sticks, and subsiding with them, gets the appellation of dregs, from which the vulgar are wont to distill a Spirit of Brandy Wine:

But these dregs are not a useless matter, as they have been hitherto thought, for much TARTAR may be extracted out of them with little cost.

But he that shall search more narrowly into the thing, will find a way whereby poor and decaying Wine may be made good.

In some places, as suppose in FRANCONIA, ALSATIA, ANSTRIA, and the RHENISH TRACT, most fruitful for store of wine, these lees of wine are made nothing of, but are given to Swine and other Cattle to drink, by which means the Tartar in it so purges them, that they are soon fat; they seldom try to draw the spirit thence. Other-where, where wine is not made, especially where they fetch their wine a great way, they are much esteemed of, not only because wine may be drawn from them, but also put in small Hempen sacks, are pressed in adapted Presses, a good mixt wine flowing thence sale able to others; yet this being but a small wine, many have designated it for Vinegar, to which it best agrees; but he that knows how to restore to it what it lost in pressing, may make a palatable strong wine, not inferior to what it was at first: But this Secret be longs to another place, I shall here only treat of Vinegar and Spirit of Wine.

In HOLLAND, FRANCE, ITALY, this pressing of the dregs and preparation of Vinegar is of frequent use, and very profitable to many, who get nothing but what they squeeze out of the dregs they have of the Vintners, and convert the Wine into Vinegar; the residue of the dregs they put into Barrels, and sell it to Hatters, which being boiled in water, they thicken rough Hats with it; for Wool is brought into a little compass by hot water, and by how much the hotter that is, the Hats are made the thicker by it: For since it is the nature of Tartar to make the water wherein 'tis diluted hotter than fire, and since there is much Tartar in the dregs, by virtue of which, water acquires a more intense degree of heat, which the ignorant Hatters ascribe to the limosity of dregs, rather to be attributed to the inherent Tartar, hence they put so much dregs in every pot as they know to be needful to the constipation of the Wool.

And this is the use to which pressed Lees are put, but if they have a greater quantity than the Hatters use, sometimes they are corrupted, part turns to Worms, and afterward to a most stinking dirt fit for nothing. When they sell them to the Hatters, then the Vinegar made of the wine pressed out of them, which cost them little, brings them great profit, otherwise they would not gain much by their own Art, if they were not eased by the Hatters taking the residue.

But after what manner Vinegar may be made, I shall briefly show:

Vinegar-makers dispose many Hogsheads upon Stands a foot high from the ground, under which a pot to receive the Vinegar is set, erected in a hot place, the upper head being taken out, about the middle of the Hogshead they place two pieces of Wood across, sustaining another foraminous bottom, upon which they put the Vinous dregs, filling the Vessel even to the top, then they cover it with the head or some good square Cover, that little Air may enter; when in two or three days, by putting in your hand, sufficient heat is perceived, the wine which before was pressed out of the dregs, is to be poured to it, so that the wine may be above the dregs, so the Hogshead close covered is left, till on the opposite row another Hogshead can be filled with the like dregs and grow hot; then the Wine being drawn out of the first hogshead, is poured into this, till the dregs in the first hogshead begin again to be hot, which are again to be macerated in the first Wine. This mutual effusion and depletion of the Wine is to be repeated till a just acidity follows, which may be done in the space of two or three weeks, according to the heat of the Air; which will be good and permanent Vinegar, to be kept for sale in another Hogshead. Presently the first dregs being hot, fresh wine is to be poured to them, doing all things as before, till this wine also shall become good Vinegar. If you have many Hogsheads, you may make Vinegar all Summer only, but if you will also work in Winter, the place may be made hot by the help of a Furnace.

After this manner in FRANCE and HOLLAND many Families are maintained and enriched, by the dregs of Wine pressed and converted to Vinegar.

And now I shall produce my Invention, a good Gift of God, as I take it, to good men; 'Tis certain, that in Viniferous Countries, where much dregs are brought for a small price, they may obtain a profit to themselves not to be despised; for which (next to God) they are beholden to me, who by unwearied toil have found this Science, and let them show their gratitude towards Needy Persons, lest they incur the displeasure of God, and so turn this Blessing into a Curse.

First of all, let there be a good Press, with a long Beam, from which hang weights, let there be put in small bags at least 5 or 6 Firkins of Lees, and them soundly pressed.

The wine that comes out by pressure, you may turn to Vinegar, either by the help of other Lees, or some other method; pour as much of the thick in the sacks with a due proportion of water into a

Still, let it not boil in stilling, the Head being on, apply the refrigeratory, manage the distillation as otherwise Spirit of Wine is wont to be drawn, keep the Spirit proceeding thence till there is a good quantity of it, which you ought to rectify or clarify by several distillations: (though this Spirit is not abundant in the dry dregs, yet there is sufficient to pay for the making of the Vinegar and Tartar) when there comes an insipid water, take out the fire, open the Pipe or Cock coming from the Still, let the water with the dregs run through the Channel into a great sack put to the Press, when 'tis close tied, press it, that the moist may be separated from the dry: The Liquor cooling in a peculiar Vessel, and congealed into Tartar, will stick to the sides, the rest falls down like sand: Take out the dregs remaining in the sack, dry them in the heat of the Sun, button them under a Still as wood, and you will have very sharp ashes, commonly called clavellated ashes, useful for DYERS, the price of which (at least) exceeds the pressing and preparation of Tartar. When you have finished one distillation with the Press, fill the empty Distillatory with fresh water and dregs, as before, and while you draw spirit thence, press the first dregs, and continually put the distillation to the same Press, which coming out hot, don't put it to the former, but give proper Vessels to each distillation. The day after, put the first water from which the Tartar went to the bottom, to the third distillation, instead of fresh water, that is, by dissolving and distilling new dregs in it, as I have taught: Let the faeces of the fourth distillation be poured in the second water, which hath let fall its Tartar, and so afterwards by which Compendium you will make no waste of Tartar, but what remains in the water, may take increase from the new dregs to which it is poured, and it will give its Gain: But if you will, you may animate that acid water from which the Tartar is separated, and with a small charge, by Grain and Fruits of Trees, give it life and soul again, that it may again be made Wine, not unlike the first pressed from the Lees, after the same method it is turned into Vinegar, (because 'tis already sharp, the life or soul only that preserve from Corruption being wanting) the way of proceeding will bring great Gain to him that understands it.

If you know not the administration, do as I have ordered, and instead of common water, dissolve the hard dregs in this acid, and you will have good Tartar; but that languid water may also be adhibited to other uses, which I shall set down in the end of this Book: This do, that the faeces be not burnt by too great a fire, nor the spirit stink and the still be spoiled, which you may pre vent if you anoint the bottom with Lard before you put in the heavy water, and draw off the Spirit. Here I might show a Compendium whereby the dregs, though dry, should not be in danger of burning, but it would be long and tedious to the Reader. But if you mind diligently what I have written you will get enough.

When you have gotten a good quantity of Tartar from the Lees, not neat to outward appearance, but very like dirt or sand, purge it yet more, that either it may be reduced to a very fair Crystal, or till it cleaves in great Fragments to the sides of the Vessel, which you may effect by the following Way. Fill a VESICA with limpid water, almost to the top, put a little granulated Tartar into that water, soundly boiling to dissolve, putting a stick to the bottom, you may see when it is dissolved, add more by degrees, till no more will melt in the water, which you may thus find out:

Fill a Pewter Saucer with this Liquor, if it contracts a little skin or cream, it shows that if has enough Tartar to produce Crystals, but if not, add more Tartar, till you see this sign in the pot; while it boils, add the scum taken off with a perforated Brass Ladle, to the dry Lees, which by reason of the Tartar adhering, when burnt, conduces to the clavellated ashes; when the boiling has drawn Tartar enough, let it be let out through the Pipe of the Still into a clean Receiver, where while the water cools. The Crystals will incrustate on the sides; apply the rest of the water to the same uses, you will do better; if you shall change it into Wine or Vinegar, let the Crystalline roughness remain in the Vessel, for if you add a new boiling, greater and fairer sorts of Crystals will be made, because by how much the thicker this Tarter is, so much the dearer 'tis held; and thus you may conciliate thickness to your Crystals. When it shall seem enough, pour the rest of the Solutions into another Vessel, in which the first Tartar has concreted: let it dry of itself, afterwards strike the sides of it with a great blow, and great Fragments of Tartar will fall down, which you may sell for profit, put the vessel to the same labor again.

This is the exact description of producing good Vinegar and Tartar necessary, for many Arts, from the basest dregs, and of justly getting much profit with little pains.

Now I proceed to make the Wine which some call Adust, or Spirit of Wine, of the Lees of Wine, without any Costs.

After the desiccated Wine is poured out, put the faeces into the Still, add as much water as the dissolution of the inherent Tartar requires, kindle a fire, stir it at the bottom with a wooden Ladle, that it may not burn to; when you can no longer bear the heat of the water with your hand, and the spirits break forth, take out the Ladle, put on the Head, and distill till you perceive no spirit, but mere Phlegm; the remaining Liquor draw out into a sack close tied, work it in the Press, and you will have spirit of wine by distillation, and also Tartar by expression, concretion, crystallization, with double increase; as I shall show more clearly by and by, that before you begin to work, you may be sure, how much you hope to gain thereby, lest you labor in vain, maturely consult that you may discern by a premised computation what is to be done.

For instance, If five Firkins of Lees cost half a Dollar each, five Firkins will yield two R. Dollars and an half; from these I press out two Firkins of Wine, two Firkins remain in the sacks, one Firkin is supposed to waste in the transfusion, when yet a Vessel seldom soaks up four quarts: I turn the two Firkins of Wine into Vinegar; the two Firkins of thick Lees being diluted with Water, I draw Spirit of Wine and press out the Tartar; this Wine drawn by the Fire, together with the dregs left of the Tartar, being reduced by burning into clavellated ashes, do equal the Charge of making the Vinegar & Tartar, but two firkins of dregs afford from 40 to 100 pounds of Tartar, as the Wine is sweeter or more acid, (though the harsh affords more Tartar than the sweet) but if they yield but 40 pounds, this Tartar refunds the price of the Lees;

the spirituous Liquor and clavellated Ashes, recompense the Wood, and the two Firkins of pressed wine you have over-plus to be turned into Vinegar.

Or if you buy two Firkins of Lees for one H. Dollar, first you may draw off the spirit; after that, you may press out about ten pounds of Tartar, both together saleable at one Dollar, as much as the Lees cost; all the Liquor of the Spirit will be clear gain, by an egregious short labor to grow Rich. Suppose a Firkin to cost a Dollar, and to yield not above 9 pound of Tartar, each of which is worth two Bazeins; though some Lees are better than others, yet seldom any are found so poor, that do not afford their value in Tartar; even the RHENISH, FRANCONIAN, and AUSTRIAN, one Firkin yielding one Dollar, oftentimes affords fifty pound of Tartar, or more. Besides, we must note, if the Lees are dear, Tartar and burning spirit of wine can't be cheap, because the dearness of them depends on the scarcity of wine: But if the Lees are dear, the Vinegar, Spirit, and Tartar cheap, (which yet was never known) so that the quantity of Tartar will not yield the price of the Lees; neither also can you have the Vinegar and spirituous wine GRATIS; yet you shall have this profit above others, living on Vinegar and burning wines only, that the Tartar and clavellated Ashes collected from the refuse of the eliquated wine, by others rejected, you may get an help, that you may sell the spirit of wine, that cost you little, for more than otherwise. If therefore he gains by only pressing wine, or distillation of the spirit, how much rather you that acquire both Vinegar and spirit of wine from those dregs for a gratuity?

One thing more I add for confirmation: If 8 Firkins of Lees, that cost 4 Dollars, make 4 or 6 Firkins of vinegar, or one of burning wine or brandy, that vinegar or brandy yields 10 or 12 Dollars:

The pressed or distilled dregs, if they give fifty or sixty pound of Tartar, they equalize the first price of the dregs; but if there be a less quantity of Tartar, the vinegar or spirit will stand you in but little, which being made better, or sold cheaper than ordinary, you (the seller) will allure buyers before others.

Some man may say, the Art indeed has been unheard of, and i.e. gainful, but does not agree with every age and condition; for all don't understand Merchandising, but rather some abhor it, seeing the Scriptures completely testify, that the fraud between the buyer and seller sticks as fast as a Pin between the stones of a wall; to whom I answer, that to every Christian an honest Occupation is permitted by Divine Providence for the sustentation of him and his, but he need not burden

his Neighbor in any case. For that it is lawful to
endeavor a livelihood this way, or that, is at
tested to by a laudable example, in both the Old and
New Testaments.

This is the true way of pressing wine from
Lees, and reducing it to Vinegar, and drawing out
Brandy and Tartar, and incinerating the refuse.

It remains that I set down the uses to which
the rest of the water from the Tartar may be
adhibited with profit; but we must know by the way,
that the sourness in that water is nothing but
Tartar dissolved, which does not take a Crystalline
form at the sides of the Vessel; therefore this acid
water performs whatsoever is effected by Tartar
diluted in common water.

Since there is no man but knows how to dissolve
Copper in tartarized water, without the help of
salt, as in the dealbation of ordinary Coins and
Silver Vessels among the silver-workers may be seen,
who take away the Copper sticking in the superficies
of money of silver-vessels by boiling, the silver
appearing as if it had nothing of copper mixed with
it; to this work the aforesaid water may be very
profitably applied; and seeing that a great quantity
of it remains in the working of the Lees, it can
scarcely be better employed than by the following
method. If the viler Marcasites of Copper,
everywhere obvious, and which cannot otherwise be

melted with profit, as also the SCORIA or dross of Copper, which is thrown away, be boiled in this water, it will attract the Copper to itself; cast in Filings of Iron, and they will be exalted into Copper, and the extracted Copper will be precipitated from the water by the addition of the Iron. This gradation of Iron into Copper being done, which in outward appearance will be like a dirty Mud, it is to be edulcorated with clear water, and then thoroughly melted by the force of fire, which will yield a very fair Copper, no way coming short of other Copper in goodness. Which Labor is very profitable, because a great quantity of tartarized water proceeds from this Work; but that it may the more easily act upon the Copper, in extracting it from the stone or Marcasite, a little salt is to be added in the boiling.

But if the Marcasite or Copprious-stone abound with too much sulphur, which yet seldom happens in the more ignoble Marcasites, they must first be torrefied, that the sulphur may be burnt, then to be ground, and boiled with the tartarized water, and precipitated with Iron; which way more Copper is acquired than was in the Copprious-stone itself, because while it is precipitated, part of the Iron is exalted to Copper, part turns to Vitriol unlike to the natural in hardness; for a green solution remains, whereby Cloth and Yarn are made as black as with common Vitriol; also it affords the best Black for Shoemakers to color their Leather, and is naturally adapted with Oak to conciliate a black color to Wood.

If this vitriolated Water be boiled in an Iron Pot till it be dry, and the residue melted in a strong fire, you will obtain a most tenacious or tough Iron of wonderful virtue, if not tube preferred to good Copper, at least to be compared with it. And so the common Copprice—stone of Iron is better than the filings or scales of Iron, if with the Lees of wine it be reduced by the spirit of the world into small Balls, when you melt them, a very ductile and pure Iron will come out, fit for more uses than common Iron; which labors if any will enter upon, they will reap fruits not to be despised.

It would not be beside the matter to annex the manner how you might use both your Tartar and Ashes at home, neither that you might be forced to sell them abroad, but it would be too long, and rather hurtful than profitable to the present business. This I think to treat of farther another time, by the Grace of God.

Here finishes the Work of TARTAR.

BOOK OF FIRES

In which is Treated of strange and hitherto unknown FIRES, to what end they serve, and what great Profit may by them accrue to Mankind.

Beloved Reader,

That which hath moved me to write this small Treatise of SECRET FIRES, with their great Operations, hath been only this, viz. Because GOD ALMIGHTY hath commanded us not to hide our Talent in the Earth, (like the Unprofitable Servant) but to improve and augment it, as we may see in the 16 Chap. of ST. LUKE. Seeing therefore that the most merciful GOD, hath revealed those high Secrets to me, in this my great Age, makes me judge that such great gifts of GOD, have not been discovered for my sake only, who by reason of my great Age have, as it were one foot already in the Grave; but that I might make known such wonderful Works to the World.

Of the first Fire, and its Profitable use.

How to make a Lamp, which being closed up in a Glass, may be made to burn continually by its own virtue, and give light without any other help.

Of the benefit of this Lamp.

Such a Lamp may serve all those who love to see light, as the Image of GOD; and may also be very available to those who by reason of continual Weakness are forced to lie much upon their Beds; which they may keep continually burning in their Bed-Chamber, not only because of its clear shining light, which doth neither smoke nor scent, like all Other Combustible Lights, which scent is very pre-judicial not only to the Sick, but also to the Healthy, because such smokes offend the Heart and Brain. On the contrary, this Royal Fire is pleasant, strengthening, and refreshing to the Heart and Brain, and its Cotton or Wick is made of incombustible matter, and casts a sweet flavor all over the Room, This secret Fire might also serve the Hermetic Physicians to prepare a Universal Medicine; especially if the sweet refreshing Odor, which proceeds from it in Burning, be received in a Philosophic Receiver, and administered to their

Patients, as a great Cordial, renovating, and restorative Medicine.

Although this secret Fire, be of far greater advantage than what hath been mentioned, yet for certain reasons its further use must be concealed. This Fire is appropriated to the Elemental Air, seeing it hath its own Air in itself, by reason of which it burns in a Glass well closed, and if touched by the least Air, it extinguishes and goes out, of which you may see more in the Fifth Part of my PHARMACOPEIA SPAGYRICA.

There is another unknown Fire, which is appropriated to the Earth, because it resembles a black Earth, and being shut up in a close Vessel of Glass, will keep Its Fire perhaps to the end of the World, provided there cometh no Air into the Vessel; but if one would have it a visible Fire, the Glass must be broken that the Air may come to it, so the matter will presently enkindle itself by the Air, and in a short time burns to a coal, yet without flame or smoke, and at length is consumed, and leaves heavy Ashes behind.

Of the Use of this Earthy Fire.

Although it may seem to many, that this Fire is not fit for anything, yet I affirm it to be fit for great and weighty matters, which are not fit to

93

mention, much less to make known, by reason of the abuse which may be made of it; this Fire being ten times more forcible than Gun-powder. Moreover, it also plays its part in the Melioration of Metals, for when this black Earth-fire is fluxed in a Crucible, then poured out, and exposed to the Air, it begins to move, and grow as it were living, so that one may see how forcibly it attracts the Air, fills itself with puffing Bladders like an incensed Toad, continuing such a motion for the space of three or four days, then the matter falls asunder, and there flows out of it a red fiery juice, the remainder is dead and unprofitable.

With this red juice, wonderful things are to be done in Alchemy, not necessary to be mentioned in this place, because it is my desire only to make known the Wonders of God. But he that will know more of it, let him read the Fifth Part of my PHARMACOPEIA SPAGYRICA, treating of the Stone of the Wise, where he will find how the blood of this Toad doth in few hours draw the Color of Gold, so that the Body remains white, therefore it may be supposed to be the Chalybs of SENDIVOGIUS.

This Fire is also made of Steel, for when it is touched with a wet finger, sparks fly from it, as from a red-hot Iron which a Smith is forging upon an Anvil. But enough at this time of this wonderful Earth-fire till another opportunity. RIPLEY seems to have known this Fire, and of it to have prepared his Universal Tincture; his Dream sounds thus, I have seen a red Toad drink so much and so long of the juice of the Grape, till his inwards did burst asunder, etc.

The third Secret Fire is appropriated to the Elementary water, because it softens and dissolves all hard Subjects; it also radically joins the different nature of Metals, into wonderful tinging Stones. Seeing that this Water-fire is of great virtue to the Metals, both particularly and universally, I am resolved (for brevity sake) to make known its virtue in a few cases, the rest will be known by the diligent. First, this wet Fire hath such great virtue, that it can dissolve radically the most fixed Gold, within the space of twelve hours, as also to unite it with LUNA, to volatilize and render it irreducible. Which volatile SOL and LUNE, is no more like unto SOL and LUNE, but in color, easy flux, form and weight, appears like unto black SATURN; therefore the ancient Philosophers, because of its form and similitude, have called it the black Lead of the Wise, and because of its volatility the MERCURY of Philosophers; which MERCURY or black Lead of the Wise may be also per Se, brought Into a true Tincture.

Therefore they have all erred, who have taken ANTIMONY for the true black Lead of the Wise. He that will know my sailings in this matter, let him read the Third Part of my PHARMACOPEIA SPAGYRICA, where he will find Wonders, especially how I have cleansed it to the highest degree, and concentrated it into a white Mercurial fusible Stone, which doth show such virtues in Medicine as are to be admired. This white Stone hath such virtue, that it cures all diseases, if it be held every morning only a while in the mouth; of which I have treated more clearly in the Fifth Part of my PHARMACOPEIA SPAGYRICA. This true black Lead, which by this humid Fire, is made of Common SOL and LUNE, hath no equality with ANTIMONY nor other common Lead.

What else is to be done in the particular Melioration of Metals, doth not belong to this place, but shall, by the help of God, be declared elsewhere.

The moist Fire hath power to extract all SOL and VENUS out of LUNE, in few hours, if the LUNE be first granulated, and then put into it, without dissolving the LUNE, so that the LUNE remains almost as fine as from the Test. The extracted SOL and VENUS may easily be separated from the Water, each by itself, by precipitation, insomuch that the Water retains its virtue, and there is no loss either of the SOL, LUNE, or VENUS. In like manner the allay of SOL and LUNE, by VENUS, may be extracted with ease, in great plenty, and for little charge. Which Secret would be of great use to Refiners, or such as have much Gold and Silver to be parted, as in Mines, & etc.

It is sufficiently known what great labor and charge is required by the common way of parting Gold, and Silver, and Copper, & etc. therefore needs not to be repeated.

Here follows a shorter and easier way.

First, if my impure Gold be In great lumps, I granulate it; but if it be Jewels, Rings, or Chains, I only neal them, and therewith fill a parting glass to the third part, then I pour thereon of my cheap and new invented MENSTRUUM, so much as may swim above it a fingers breadth, put a Limbic upon it and set it in a Sand furnace, give fire by degrees till the Liquor boil, and be all gone over the Helm, to a dry Salt. So hath the AQUA FORTIS in the time of digesting drawn to itself all the allay of LUNE, VENUS, or other Impurity, out of the Gold, in form of a green Powder, which remains with the Gold in the Glass. When it is cold, I take it out, of the Sand, and put in warm Salt water, to dissolve the extracted LUNE and VENUS, that it may be poured out of the Glass; which done, I boil it in a Copper Kettle, so the Salt-water extracts the LUNE and VENUS from the SOL, and makes it fine. I decant the green and thick water from the SOL, and put to it in the Kettle fresh Salt water, and boil it to extract the rest of the LUNE and VENUS from the SOL. But if this second Water should yet be green and thick, that is a sign that there yet remains more LUNE and VENUS with the SOL; therefore more fresh Salt-water is to be added to the SOL, as before. After boiling, put all the green Waters together, and at last pour upon the Gold only common sweet water, with which boil the Gold to extract the remaining saltiness from it, which will leave it fine and shining in the Kettle, as fine Gold used to be, which is to be set upon a small fire to dry, this to be kept as fine Gold. The green Water Is to be filtered, so the LUNE will remain behind in the filter, which is to be dried, and mixed with the common fluxing Powder of SULPHUR and NITRE, and being melted will be reduced to the finest LUNE. Boil the green filtered Water in an Iron Pot, so all the VENUS will settle to the Pot bottom, which is to be edulcorated with Water, and being mixed with the above named fluxing Powder, and reduced becomes the finest VENUS that can be seen with eyes. By this means I do not only find all my SQL, LUNE, and VENUS, but also an increase of the SQL, the reason of which increase proceeds from my AQUA FORTIS, which is a graduating Water, and of a meliorating nature and property; which increase indeed is not over great, yet it pays for all the charge of Coals and Labor. This work may be performed in the space of twelve hours without any loss 'of SOL or LUNE; which differs greatly from the tedious vulgar way of separation, which is very laborious and costly, besides the great loss of the SQL and LUNE. So that ten Marks may be easier separated by my way, than one Mark by the common way.

One great Secret more, above others, which for brevity sake cannot be all inserted here, is this. It is well known to all Chemists, that all Vegetables, as also Animals, by addition of common Water, may be brought to fermentation, and according to every subject, a SPIRITUS ARDENS, of great use in Physick, may be distilled. But how to make such a subtle Spirit out of Metals, I never read in any Author, nor heard of. But I have found out a way by which great matters may be done, which cannot be mentioned here, let others search after it as I have done, it is not good to cast Pearls before Swine. Yet that the desirous may know somewhat of an Artificial Metallic Fermentation; I say that out of certain Salts a Water may be prepared, which when it is put upon a compact Metal, that the same by a certain property in the Water, begins to swell up and ferment, like to the fermentation of Wine, Beer, or other vegetable Drinks; and after fermentation, by distilling in Balneo yields an exceeding subtle penetrating Spirit, strong, sweet, and volatile, according to the nature and property of the Metal;

which may also by rectification be concentrated and subtilized, like that which is drawn from Wine or other Vegetables, without any Corrosion; so that such a Spirit may be put into the eye without the least hurt, or sense of sharpness, (See the Sixth Part of the PHARMACOPEIA SPAGYRICA) Therefore it may easily be concluded, that such a Metallic Spirit may effectually be used in the most difficult diseases. But those Spirits are not only highly profitable In Medicine, but cannot also fail to show their effects upon Metals; for I have already experienced thus much, that the Spirit of SOL doth in few days graduate Common MERCURY into SOL, in like manner doth the Spirit of LUNE and JUPITER graduate MERCURY into LUNE; other Spirits of Metals I have not yet tried. And although many may think to force such a Spirit out of Metals as is drawn out of Vitriol, yet they err, because every Vitriol distilled per se besides the Phlegm, acid Spirit, and heavy Oil, yields also a volatile Spirit, which may be very much subtilized by rectification, but it hath no comparison at all, nor likeness with my new invented volatile humid Spirit of Metals; because the above mentioned volatile Spirit of Vitriol, consists only of a subtle salt Spirit and spiritual Sulphur, which may sufficiently be demonstrated. For when such a subtle salt Spirit of Vitriol stands, a while in a Glass close stopped, the volatile Sulphur of the Vitriol sublimes itself to the top of the Glass, and congeals itself in the form of common yellow Sulphur, and the humidity remains in the Glass like an insipid Water, without smell or taste. Therefore the volatile Spirit of Vitriol now days prepared, is to be esteemed for nothing else but a volatile Spirit of Sulphur. In opposition to which my Metallic Spirits, especially those of SOL or LUNE have no combustible matter in them. Also the Spirit of Vitriol is yet corrosive, which the Metallic Spirits are not.

Therefore let no man think that my invented Spirits of Metals are of the same nature with the sulphureous Spirit of Vitriol, Allom, or common Salt, or the like, but let it be esteemed as a new-in vented high Secret, the like of which was never yet known to the World. He to whom God shall please to reveal how these Spirits of Metals may again be brought into fixed Bodies; such a One may well esteem himself happy. Seeing that such a Spirit being yet fugitive, can mortify running MERCURY, and transmute it into SOL. Also know, that our secret Salt water hath power to bring other Subjects into fermentation: As for example; if I was willing to bring Tartar again Into fermentation, to try what Spirit it would yield, I dissolve it only in fair water, and put to it a little of our Universal ferment, so will the Tartar presently begin to ferment and work, and yields a wonderful volatile Spirit, which is not to be had by any other way. Here I will add only one thing by way of history; namely, what happened unto me once, When I joined a little of this ferment with some Urine; that the same did presently begin to ferment, and yielded a most dreadful scent, so that I was necessitated to carry the Urine out of the room, the smell was so offensive, that it enfeebled my Heart and Brain; and In the distilling, yielded a scent far worse than the putrefying of any dead carcasses of beasts, at last I forced over all the un pleasant Phlegm, took the CAPUT MORTUUM out of the Glass, and distilled out of it a strong Corrosive Spirit, with which I did dissolve SOL and LUNE. But what is further to be expected from it, time will show me. There came also with the strong salt Spirit, a red Oil of Urine, not strong.

To enlighten the former discourse, there remains yet something to be said, namely this: That the before mentioned Spirit of Metals may be handled in all respects like the burning Spirit of Vegetables, which is distilled from Vegetables fermented; in rectifying, the most subtle Spirit cometh over first, and the unprofitable

Phlegm remains behind; and the oftener this rectification is per-formed, it is the more pure, subtle, strong and pleasant, so that the virtues of all other Vegetables, some of this Spirit being poured upon them and digested, may be extracted, and separated from their gross faces, by separating the Spirit of Wine in Balneo, so the virtue of the Herb, Root, or Flower, will remain in the Glass; of which one only drop hath more virtue In Medicine, than a whole handful of the herb from which it was extracted; yet the Spirit remains good, and is always proper for the like occasions. But if one desires no extract, but only a Spirit of Vegetables, strong and forcible in its operation, then must the Spirit be put upon fresh well scented cordial Herbs, Roots, Flowers or Fruits, digesting them together, then abstracting the Spirit again, so will it be much stronger, subtler, and of a more refreshing smell than before. Thus in like manner can we do with our Spirit of Metals, and make it as strong as we will, by putting the same upon fresh Metals, and letting them ferment together, then abstracting, and so fermenting and abstracting, whereby It always grows stronger and stronger, so that afterwards greater things may be done with it both in Medicine and Alchemy. If then the simple Spirit of Metals doth wonders in Medicine and Alchemy, what will not this doe, that hath three or four times its strength? Consider of it, enough to the wise. The Spirit of Metals thus made we esteem for the true AQUA VITAE or VINUM SALATIS of the Ancients, and no other made of Vegetables. Take notice also by the way, that our Ferment doth not lose its virtue in distilling, but serves always for the like Works, to wit, the Fermentation of Metals.

Seeing that our Ferment is a volatile Salt, after Fermentation the subtle Spirit is abstracted, and the Phlegm received by itself, so at last rises our Ferment with a stronger Fire, and leaves the mortified Metal as unprofitable faeces. So that the Ferment being once had, it may be often used, yet it is somewhat diminished every time by sticking to the glasses, yet not being costly, the loss may be easily born, so that there need never be want of it. This is indeed a wonderful], subtle metallic Wine, Of which I had once a Vision about thirty years ago (with these words: EX FECE, DE FECE, DEHES (?) CARPERE VINUM TUUM,) yet I knew not the signification to this time. I have had much thoughtfulness about it, yet all in vain, until God was lately pleased to show me the way of Preparing such Wine.

If I should write all things which may be done with my new-in vented Fermentation out of Vegetables, Animals and Metals, I should have work enough for a great Volume, and as it were bring forth a new World: But let this suffice at this time.

Next follows the fourth Secret Fire of the Wise, to perfect a Universal Medicine both for Men and Metals.

That the ancient Philosophers did make their Universal Medicine, not out of one Subject only, is clearly to be seen out of their Writings. And as their Subject was divers, so have they also used several Fires to decoct their Subject; for some Subjects, as fixed SQL and LUNE, they have made volatile by their humid Fire, and again made the volatile fixed by means of the known fire of Wood and Coals. Therefore have they so positively affirmed that without common Gold and Silver, no true Medicine for the Metals can be made. And this, by reason that other Subjects were yet unknown to them; and they knowing no better, thought theirs to be the only way by which the Universal Medicine could be made. But others have wholly rejected SQL and LUNE, saying they are dead, and that one should beware of them; but that their SOL and LUNE were living, and that no commonly known fire could ripen them, that a far other fire is required for that purpose; these and the like contrary Writings (although they are both true) have confounded the Searchers, so that one esteems one Writer and another, another, and yet are always doubtful, after which Author they should work. PONTANUS confesses that he erred two hundred times, although he wrought

in the true matter, and should never have understood
if he had not read the old Philosopher ARTEPHIUS,
and of him learned the Secret Fire. He describes
also the Secret Fire in the same manner, which
ARTEPHIUS doth, to wit, thus: OUR FIRE IS NO COMMON
FIRE, IT IS ARTIFICIAL TO FIND OUT, IT DISSOLVES,
DESTROYS AND PENETRATES ALL THINGS; IT IS EQUAL,
CONTINUAL, AND CONSTANT, BURNING WITH OUR GLASS, AND
NOT WITHOUT, & etc. Our Subject is also no common
SOL or LUNE, but as yet an unripe Mineral, on which
Nature hath but began to operate. All they who have
wrought in such unripe SOL and LUNE have also of
necessity known this Secret Fire, Or else they have
failed, and could accomplish nothing. Some have
known the volatile Gold, as also the Fire belonging
to it, and yet could not come to work with it; the
reason was, because they as yet wanted the
instrument belonging to it, because if a volatile
matter be to be ripened in an open Glass with a
flaming fire, it is impossible but that the volatile
MERCURY should fly away; and If a Receiver were
applied to receive the volatile Spirit, then by that
means the Fire would be stifled and go out. Here is
good counsel at a dear rate, but the discovery of
the secret Vessel makes all cheap again; for without
the knowledge of this Vessel, little is to be done.
Is it to be a Universal Work, so must it also agree
also with the Universal Work of this world? We see

daily how that the warm Sun causing its beams to descend downwards upon the Globe of the Earth, causes all Vegetables to grow and multi ply, and not only Vegetables, but also Animals and Minerals, as PLATO witnesses in these words: THE SUN AND MAN GENERATE MEN. So that the oftener the beams of the Sun doth draw up the moisture of the Earth, and let it fall down again in Rain, by so much the more it makes the dry Earth fruitful, by which means the Fruits of the Earth sprout forth faster, ripen, and multiply: because the moisture when drawn up in the day-time by the Sun beams, always leaves behind it its fruitful, Salt, which it had attracted from the Air; in the bowels of the Earth, and is still again impregnated with the influences of the Heavens; for without the Air, nothing can live, increase, or multiply, the which HERMES TRISMEGISTUS hath very well given to understand, when he said all which is above, is also below, by which wonderful], things are brought forth; the Sun is its Father, the Moon its Mother, the Wind carries it in its belly, the Earth hath conceived it and brought it forth, and is its nurse, by continual ascending and descending it hath obtained its force, and its virtue is complete, when it is trans changed into Earth, & etc. Here HERMES gives sufficiently to understand, that the volatile matter of the Stone is elevated from the bottom of the Vessel to the middle region of the

Air, and must again descend to the Earth, that by its continual ascending and descending it may be made to live and increase, as it is in the great World. But how to make this ascension and descention as it is and ought to be done in the Philosophic work, is not permitted to be divulged; but it is more than enough to discover and assure that in such a secret Furnace and hidden Fire, the volatile Gold of the Wise may be ripened into a living Medicine, AMEN.

Here should come in a small Treatise, entitled, THE EXPLICATION OF SOLOMON'S WORDS: viz.

In Words, Herbs and Stones, there is great virtue. But who this SOLOMON was, is altogether uncertain; and the Treatise itself consisting of nothing but fabulous Stories of divers Events, which followed upon the speaking of certain words at divers times, and upon divers occasions, and containing nothing (as I conceive) that will be useful either for a Christian, or an Artist, I think fit to leave it out; and more especially, because this Treatise is wholly foreign to the Author's knowledge and experience, but taken upon trust (as to matter of fact) at random, and from others, which is disagreeable from all the other Writings of the Author. All that I think worth the Translating is somewhat relating to the Coloring of Glass; which take as follows.

With one pound of pure Glass, mix about a dram of the Ashes of Copper, (or CROCUS VENERIS) and melt them together in a strong Fire, and you'll find a green Glass, resembling the ONYX-STONE. In like manner also a whole pound of Glass is made like to the fairest SAPPHIRE, if a dram of COBOLT or BISMUTH, melted, be added to the SCORIAE or Dross. Likewise one dram of the Rust of Iron changes a whole pound of Glass in a yellow Stone, emulating a HYACINTH. If one shall melt a dram of the Tincture of Gold, or of the Philosophers Mercury, extracted from Gold and Silver, with a pound of Glass, that Glass becomes very red, like a RUBY in color, as I myself have found. But if any shall be so fortunate as to make this Glass hard, as well as of a beautiful, color, he will need to take but little pains for his living.

There is a short Discourse adjoined to this, called, THE QUINTESSENCE OF METALS; but whatsoever is mentioned in that, is contained in the Fifth Part of the PHARMACOPOEIA SPAGYRICA, Book of Fires, Book of Dialogues, & etc.

The End of the Second Part.

A SHORT BOOK OF DIALOGUES

OR,

(Certain) Colloquies of some Studious Searchers after the Hermetic Medicine and Universal Tincture.

Written for the Sake of the Lovers of Hermetic Philosophy.

THE PREFACE to the Well-minded READER.

I was formerly minded never to have published these Three Dialogues, but only to have made some of my good Friends, and such as had well deserved at my hands, here and there, partakers of the same. And upon this Account I permitted some (of them) to Copy them out, but they abusing that Courtesy (of mine) whereby they received them, did make others of their own Friends too, enjoyers of the same, contrary to my Will and Intention; and so it happened, that they became Common, and being on this wise often Copied out, there crept in amongst them (as indeed usually falls out in such cases) abundance of Faults and Errors, and the sense (and true meaning) of my Words were construed in the worse Part. Which thing when I perceived, that it would more disadvantage than profit me (especially seeing, that such a work (thus copied amiss) did nevertheless pass under my Name, and was adjudged by others, as really mine) I deemed it, expedient, of two Evils to choose the least, and to have regard to mine own good repute, and to publish it in mine own Name. But yet, not with an intent of getting myself some eminent Fame, as if I were wiser than others, and to have it thought, that I had more knowledge and experience than many others have; but rather, that the incredible Works of the omnipotent God, and his great Wonders, might be laid open and made known, to the infinite Glory of his Name. In the sitting down of which, I do produce only such things, as myself have wrought with mine own hands, and can even yet demonstrate by a certain and undoubted Operation, (by Gods help) at any time.

But yet, I would not have any one thus to understand me, as if I had already wholly and completely finished the whole Operation, and had advanced it to a due, and thoroughly perfect end, No! I cannot arrogate to myself by any means, any such matter. Thus much I only affirm, that if any one shall (in his Operation) follow the bare literal Description of these Labors, he will without any Error arrive, so far as I myself am already come, but yet with this Proviso, that he knows the true Salt of the Philosophers, and the use thereof; And as for what remains, (unfinished) I commit unto God to bestow a prosperous Success:

And this one thing I entreat, that every Body accept of the things I have here written, with the same mind I wrote them, and that he take in good part my sincere Endeavors of deserving well at his hands.

The Explication of the annexed Figure belonging to this Treatise, noted with these Words: INDE DIALOGUS.

In the four principal Points of the Circle (supposing two transverse lines were drawn, through the Center, to the outward Circumference) are placed the Characters of SOL, SULPHUR, LUNE, and SALT.

Round the outmost Circle, are placed these Words: Conjoin in one, SOL, SALT, SULPHUR, LUNE.

About the next Circle: And thou hast as great a Treasure, as Heaven can give thee.

Within the Third Circle: The Philosophers Function is of Contraries, the Conjunction.

About the next Circle: The Concentration of Homogeneals, is the Separation of Heterogeneals.

Within the inmost Circle: SOL, SALT, SULPHUR, LUNE.

The First Dialogue

The First Dialogue, or Conference, betwixt two Lovers of Hermetic Medicine, deciphered by the Letters, A. and B. the last of which hath had a prosperous Success on his Labors, the other not, and therefore craves of this last (viz. B.) a Manuduction to the Work, whereby he is rendered Master of his desire.

B. A good health to you, my Friend! What's the matter with you now that you are so sad, and even laden with Cogitations, and mumble to yourself about I know not what?

A. Oh, my Friend! I wish you the like very heartily; and am glad that you come so very seasonably, and at such a time, as I was just thinking on you, and most earnestly wishing your approach; Witness your own Writings, which I do here turn over with my hands and my mind, but yet they are so very obscure, that I cannot worm myself (as I may say) out of them, (or understand them) though I apply the utmost of my Endeavors to understand them. I have likewise read over and over again, the Writings of other eminent and belief-deserving Philosophers; still hoping, that I should yet at length attain to the know-. ledge of the Truth: But alas, (the more's my grief) all that I find is only this viz, that I hold in my hands the slippery Tail of a slippery smooth Serpent, (or Eel) which every now and then slips out of my hands, and doth more and more defile me. I have therefore resolutely determined with myself, that, unless God doth shortly send me some good Friend, who may lead poor me out of such a notable Labyrinth, I will throw all my Books, all my Instruments, and all such matters which I have bestowed so much time about, in vain, and lost so much by, into the Fire; and Sacrifice them unto VULCAN, that so I may be rid of the tediousness of my fruitless Labors, and unprofitable Cookery. But yet if you would be but so pleased, I no ways doubt, but that you might by a few words (and Directions) reduce me out of the snares of so many Erroneous paths, and hedged up ways, into the right path: For I well know, that you have bestowed your whole Age, your whole Study, and all your Labors and Endeavors, about such great Secrets, and have by the Divine assistance obtained the very Truth itself. And therefore I do most humbly beg at your hands, that you would not leave me destitute of your help, but that, according to your inbred Goodness and Courtesy, you would succor me, your Friend, with some brotherly instruction, and Manuduction. Which if you either will not or cannot do, I must even conclude, not only upon thoroughly doubting of the Truth and possibility of this Art, but withal, on a firm persuading myself, that those

Writings which are so stuffed with the Promises of golden Mountains, are nothing else but mere Old Wives Tales, and frothy Speculations of idle Men, and vain Dreams, though proceeding from Men of so great Esteem.

B. But what's this, I hear thee utter? I could never have believed you, to have been of such a broken and dejected mind. What? Would you contemn the Writings of the Philosophers, and slight them, because they are above your Capacity, and too hard for your understanding? Tis a wicked thing, to entertain such a thought, much more to utter it. I would have you, rather to persuade yourself, that you are not as yet worthy of the Secrets and Gifts of so great worth: For though a Man should torment himself with abundance of hard Labors in this World, and should afflict his Body with incessant Sweating pains, yet would he not effect ought without the Blessing of God. Do you not know that saying of PAUL; TIS NOT OF HIM THAT WILLS, NOR OF HIM THAT RUNS, BUT OF GOD ALONE THAT SHOWS MERCY. You should therefore reckon yourself amongst the number of those, that have run in vain, nor hath God injured you at all. What does not Christ say, NOT ALL THAT SAY UNTO ME, LORD, LORD, SHALL ENTER INTO THE KINGDOM OF HEAVEN, BUT THEY ONLY WHO DO THE WILL OF MY FATHER. Examine now yourself, and see how the Case stands 'twixt God and you. The bestowing of such great things must proceed from God, and not from the Philosophers. The Philosopher may indeed write down the Truth, but yet it is not in his Power, to bestow upon thee the Divine Blessing, which is the very hinge on which all good things depend. Secrets of such great moment are not the Gifts of Men, but of God, who bestows them on whomsoever he pleases.

A. In good time! Is this the Comfort and Instruction, which I begged at your hands? I did not request, you to be my Father Confessor, to hear my Confession of my Deeds, but rather that you would help me, being ignorant and unskillful, by some good and profitable Manuduction and Instruction: For I well enough knew, that wicked Men are never Masters of such great Secrets, nor will I rank myself amongst them. Be pleased but to regard my suit, and only show me an entrance, whereby I may enter into the right and Kingly way: And as for praying to God, and Laboring without ceasing, leave the Care of that to me: I hope, that God will not deny his Blessing upon my Prayers and Labors.

B. Well! since I perceive you to be so thoroughly bent, with your utmost study and unwearied pressing on, after such an eminent thing as this is, I cannot but show you that way, which I myself have walked in, and that too, home to the very place which myself am come unto. Verily, I see the Promised Land afore my Eyes, and do daily view its Coasts, nor do I doubt, but that I shall shortly enter thereunto, and have the Fruitation of its most pleasant Fruits, if no impediment debar me of so great a happiness. And as concerning yourself, seeing that you are nimbler of your Feet than I am, there's no doubt but that you will arrive thereunto, even as soon as I myself. But yet, pray first declare unto me, about what things it is, that you have spent your Monies, your Labors, and your Precious time, and all to no purpose; that so I may (as much as in me lies) the more conveniently reclaim you from your Wanderings and Errors into the right way. Tis in vain for him that is sick, to expect help and succor from the Physician, if he does not show the place of his Dolor and Grief. Confession is a Medicine to him that goes astray. Confess therefore the Truth that I may hear, by what things thou hast been misled into so many Errors.

A. (Alas, Sir,) I could not reckon up all, in Order, though I should have time enough of so doing. But your own time, which is far more precious, does not permit, that it should be spent in hearing my foolish Labors. Besides too, the remembrance of so many Labors in vain, and of the loss of not only so much Time but Expenses too, causes a loathing in me, the very remembrance of which I abhor, much more to make a long rehearsal of the same. You may therefore

easily guess that by my insisting upon the bare
Letter only of the Philosophers writings, and not
understanding the sense and meaning, I have erred
from the right way, and have headlong buried myself
into so many Intricacies and Errors. I have searched
into Vegetables, Animals and Minerals; but I see
that I have not had under my hands the true Matter.
For if there does appear in any (of these Matters)
the Crows head, yet the other Colors which the
Philosophers make a description of (as the Dragons
Blood, the Peacocks Tail, Virgins Milk, Coagulum, or
Curdling, and principally that Red and Fire-abiding
SALAMANDER) did never appear (to my view). Or if
these (Signs) of SANGUIS DRACONIS, or LAC VIRGINIS
appear in sight, in some other Matter, yet
notwithstanding the other Colors, and other Signs,
which the Philosophers make mention of, did never
discover themselves (to my view). What Labors soever
I have used, and whatsoever matters I have dealt in,
I have even Labored in vain, and lost both my pains
and Expense, and never have received any good from
my laborious Operations. Hereupon I did at last even
almost thoroughly persuade myself, that it was an
impossible thing, that, out of one Matter, and by
one and the same Labor, one Color should orderly
succeed another, and become visible to the sight, by
the bare help of an external Fire, as for example,
first of all in the putrefaction, the Crows head,

then the Peacocks Tail, then the Dragons Blood, LAC VIRGINIS, COAGULATE or Cheese like Curdling, and at last the fixt SALAMANDER. But forasmuch as it appears to me, by the reading of your Writings, that you have orderly met with the sight of all those Colors in your Labors, in such manner as the Philosophers have described the same, I do firmly believe, and give Credit unto your Sayings, as unto a Man that makes Conscience of his ways, supposing, that you would not write such things unless you had wrought them with your own hands, and could even yet perform them at any time. I only beg your help in showing me the true matter, and the Key thereof, that I may so order the Business, as to cause the Visibly appearing of one Color after another, in one Glass, and by the bare help of one only Fire; if you do but thus much for me, you may be confident that I shall be the most contented Man (alive). Nor do I doubt, but that as touching the remainder, as Multiplication, Projection, and such like, I shall find out those Operations well enough afterwards, by mine own studious Search, if I can but once hit the entrance of the right, true and Kingly way.

B. (Hold a little, and) do not assume so much unto yourself, and think that the things which are so easily said, are with as much Facility done. Have you not Read in BERNHARD TREVISAN, that a certain (Friend of his) had that great Secret as well as himself, only he knew not how to multiply it, nor would BERNHARD reveal the same unto him, as having the selfsame Books, out of which the said BERNHARD got the knowledge of Multiplication, himself. But be it as you desire, and seeing you request no more from my hands at this time, but only the matter and some Key; I will satisfy your request, as far forth, as the time and occasion will at present permit.

Attend therefore with diligence to those things which I shall say unto you and such things they shall be too, as unto which you may boldly give Credence. I will not (according to the Custom of many) seduce you, nor will I reveal unto you ought else, but what I have experienced by the Labors of mine own hands: And if you follow the guidance of the bare Letter itself, you will not err, unless God will not permit you to proceed, (but) through some peculiar impediment and let in your way.

As touching the matter, which the Philosophers have made that Universal Medicament of, I find that it is not merely ONE, but DIVERS, and this is clearly evident from the Writings of the Philosophers, who openly hint unto us, that one of them used this way and matter; another, that, and yet at length became Masters of their desire notwithstanding. From whence it necessarily follows, that the different matters, of which is made one and the same thing, are not unlike in the more inward parts, but alike, though they do not appear so to be, as to their external hue. For it is a thing possible for two, three, or more things to differ much, (from each other) as to the outward form and shape, whereas notwithstanding in their inward parts, they are so agreeable, to each other, as that the self-same thing may be produced from the one as from the other. Take, for an example, of this thing, the SEEDS and ROOTS of some Herb, the which, as to the outside form, have no likeness to each other, and yet for all that, do they produce one and the same Herb, if they are implanted apart in the Earth. Just thus is it with the Metallic Buds and Stocks which are wont to sprout forth, as well from the Metallic ROOTS, as from the Metallic SEEDS, in so much, that a Tree grows up of the same Nature and Form from the Metallic ROOT, as Springs from the very SEED it se]f. Now 'tis evident, that in the Metallic Kingdom, SATURN or Lead supplies the place of the ROOT; MARS or Iron, of the Trunk or Stock: JUPITER or Tin, of the Bark, MERCURY or ARGENT VIVA, of the juice betwixt the Trunk and the Bark; VENUS or Copper, of the green Leaves; LUNE or Silver, of the white Flowers; and SOL or Gold of the ripe Fruit and Seeds. If therefore the Metal-lick plant is to be multiplied, that Multiplication cannot be more commodiously effected then by SOL and SATURN, that is, by the Seed, or by the Root of the said Tree.

Whosoever therefore desires to perform ought in this kind, he will not find any more convenient matters, then SOL or SATURN, that is, Gold or Lead. But yet I do not mean those vulgar Metals, but such, in which the Gold lies as yet immature and invisible, and which is to be made visible, fixt, mature and constant by the help of Art. So then, the self-same thing which may be discerned, above, in SOL, and appears visible to the sight, is in like manner found beneath in SATURN, in an invisible manner. And thus experience itself shows that, out of two things unlike, as to the outward shape, one and the same thing like them may be made, because their internal parts are of one and the same Nature, and this outside difference or unlikeness proceeds only from the impurity, and defect of Maturation. Out of SATURN therefore as out of an unripe and impure Gold, some good may be produced: But it must of necessity be well washed, and out of it being well washed, may the first ENS of Gold be extracted, and be fixed. But now, if out of mature Gold, you would yet educe something, it must then again first putrefy and be reduced into nothing, afore any more noble thing can proceed there out of. For it is like to the Seed of the Vegetables, which does not admit of any Multiplication of itself, unless they are first put in the Earth and consumed by Putrefaction: And this is proved, and asserted by the Testimony of

Our Lord Christ himself, who says, that except a grain of Wheat rot in the Earth, it cannot bring forth any Fruit. Certain it is therefore, and firmly true, that Gold cannot be translated into a better degree, unless it is again destroyed, and reduced into such a Body, as out of which it cannot be reduced into its former Golden Body (or Form).

A. What is it that you say, can it ever be possible, that a Metal so constant in the fire should be in such wise destroyed, as not to be reducible unto its former Body? Verily I have but small reason to boast of any great matters done by me: For I have for some years past tormented myself hitherto, about decocting and cooking of Gold: I have dissolved it in sundry sharp (and Corrosive) Waters, and have beheld its ascending with its yellow Color, by a Retort and through an Alembic, but yet I never got ought else in the Precipitation of the same, but common Gold, and which was not in the least bettered thereby. And therefore I did at last conclude with myself, as many others have done, that the common Gold could not be the matter of the Philosophic Stone, and it holds hidden within its Body, no more Tincture than it stands in need of, itself; and that therefore it has not the Faculty of tinging other white Metallic Bodies.

B. I do not at all Wonder at your falling into this Opinion. There are many Others besides you;

that are of the same mind; Nay, I myself doubted much about this very thing, viz. whether or not, Gold hides Within its INWARD parts any more of COLOR, than it shows unto us in its Outside shape. But then On the other hand, it could not seem at all likely, that such eminent Men Should publish such great Fallacies and so many Lies, merely to seduce Men, by. And Whilst I Was thus wavering in this kind of doubting, the Truth did at last (after sundry and many inquisitions) by a mere chance present itself unto me; in so much, that I am flow clearly Convinced of my Error, and am even constrained to believe, that a true Tincture tinging the.jm_ perfect Metals may be extracted out of Gold. For well may that be believed, which the Eyes see, and the hands feel.

A. I rejoice exceedingly to hear you say, that you have seen the Truth, and I hope that in time you will refresh me with a Sight thereof too.

B. Whatsoever lies in my Power to serve you by, I will not in any case deny unto you: But thus much I would you Should know, that the Splendor or brightness of the Truth itself hath Shone Upon me, but I have never as yet brought the work itself unto an end, by reason of the want of time: But yet however, I am confident and firmly persuaded, that if no impediment chance to happen, I shall bring it to its Wished end. And now seeing you are by some

125

years younger than myself, and that you have store of time and all other Conveniences I dare be Confident, that you would finish that Operation much Sooner, should I but reveal unto you those things, Which I am already arrived to the knowledge of, by the Labors of mine Own hands.

A. Proceed on, I Pray, in this your Liberality, and make me, as being a Man following after Honestly Partaker of your Happiness and I shall be everlastingly Obliged to you and Yours. And whatsoever Labor or Task is to be undergone for you; I will with a ready and willing mind undertake It; and in all things respectfully regard Your Wholesome Instructions

B. Well I trust you, and believe, that YOU will perform your Promises, by which you bind your Credit; but however you shall give me your hand, and Promise me, that you will Conceal the Art in most profound silence.

A. I will, here's my right hand, and Credit upon it.

B. Hearken then, with your utmost diligence, and with an accurate intention, receive the things which I shall speak unto you.

A. I do, and I listen attentively.

B. In the first place then, you are to know, that, if you would make any good thing out of the common Gold, you must perfectly Cast out of your mind that Opinion, Which hath hurried not a few into no small difficulties, imagining, that (by the help of some MENSTRUUM or other) the COLOR of the Gold is to be extracted out of it, and that Silver is to be tinged, with that same Golden Tincture thus extracted, and that, to the remaining white Gold, its Color may be again restored by the other lesser Metals, as MARS or ANTIMONY Copper, or Iron: Such thoughts as these you must Clearly remove out of Your mind, as being those which rob a many of their precious Time and Estates. There are several ways, by which I know how to extract the Color from Gold, but tis needless to reckon them up here by a tedious repeating of them, seeing they are not any ways profitable, but rather cause loss of Time and Goods. The main thing you are to mind is this, viz, to meditate (and enquire) by an accurate and incessant studious Search, by What means you may destroy Gold, kill it, and so compel it by Putrefaction to pro- dude to view its Internal and invisible Color, and (on the Contrary) to introvert (and hide) its external and Visible Yellowness. For Gold itself is no other thing save a mere Tincture, to the acquiring of which, there needs not any other thing save the true Key, which unlocks Gold, introverts it, and renders the invisible Color visible. Besides, neither are those to be hearkened unto, who boast of reducing Gold into its three Principles,

viz. SALT, SULPHUR and MERCURY; and of freeing those
three from all their impurities and then, of
Conjoining them again, being thus Purged, and of
Fixing them into an Universal Tincture; and such
like most impertinent trifling Processes, as these.
For they are mere idle Dreams, and can never be
accomplished, but come to just nothing, and clearly
delude the Covetous Thirstier after Gain, by their
vain dependence thereupon. Nor are there in Gold any
of those Feces, Which they prate of its being
defiled with, neither doth it admit of being severed
and dissolved into those three Principles. But put
Case it were Possible so to be, what profit, I pray,
could we hope should accrue to the said Gold by such
a fruitless Labor, whereas we see, that it is not in
the least measure bettered by such a Separation. It
remains therefore for an undoubted Truth, that Gold
neither contains any Feces, nor admits it of a
resolution into Three Principles, but that it rather
requires to be radically dissolved by a due
Putrefaction, and to be so opened or unlocked. And
farther, the Labor of such Men is likewise vain, who
Endeavour by the help of Saline, Cementations to
extract from Gold, its Soul: For though such
Cementations may sometimes succeed so well, as that
the Gold when taken out is Plainly white, yet
nevertheless such a white Gold doth as yet contain
in its own peculiar Color, the Which, a little

SULPHUR cast in upon it in Flux, doth easily restore unto it: For then that whiteness vanishes, and the Truth appears, and shows you, that it neither lost its yellow Color, nor its weight, but retained them both, in the Cementation. Nay, we have been many times deceived ourselves by these kind of Operations, and have persuaded ourselves, that we had despoiled the Gold of his Color or Tincture by the Salts, whereas it had but only attracted a certain SULPHUR out of the Salts, by which it was made White. You may give Credit unto me, for I speak experimentally and do not tell you dreaming Stories. I will instance it unto you, by an Example, Dissolve a little Gold in some AQUA REGIS, and Pour the Solution upon Powdered TARTAR, that so being Poured Upon the said TARTAR powder, it may be hid and covered over:

Put this TARTAR thus moistened with the Solution of the Gold, in a strong Crucible, the which you must cover well with a Cover, and lute it: Or rather, put it in a Cementary Pot or Vessel, Which will be better, The Vessel being placed in the Cementary Fire, the Gold will extract a peculiar SULPHUR, and become White and Brittle after its Separation from the SALTS, by being melted. And flow who is it, but would believe, that the Salts had extracted the Color of the Gold from it, whereas it is no such matter. For a little SALTPETER, or else the CINERITIUM or CUPEL, can drive away all this white Color, and restore it to its former Yellowness again; and this is, what myself have several times done and experienced with mine own hands.

A. Now again, here's a new Story I never heard of afore, who would ever have believed, but, that when they had taken their Gold (tinged with a whiter Color than Silver) out of the Cementary Vessel, it had been clearly despoiled of its Tincture? But flow seeing it is not so, there must of necessity lye hidden under such an Action as this, some other Secret and Wonder. Verily it is no trifling Matter thus to make Gold white, Without the help of the white Metals; and it is the more wonderful too, because it is not known, from whence that white Color receives its Rise: it could not get it from the AQUA REGIS, nor could it have it from the TARTAR, and this makes me still wonder the more. And therefore, pray, rid me of this doubt, and unriddle the business unto me, for 'tis not without cause, that I suppose some great Secret may be thereunder hidden.

B. Attend diligently to what I say, therefore for its impossible for you to apprehend all things at one very dash (as I may say, and at first). We will first of all treat about the Gold only, and of other Secrets afterwards in due time. But yet (by and by) I would have you observe in this place, this one thing; that as touching that SULPHUR, which made the Gold white and brittle, there must needs be a notable Friendship betwixt them, because it was so easily extracted out of the TARTAR by the Gold. And upon this Account there may be ground to suppose, that if the Gold were left lying longer in that close Cementation, that SULPHUR which rendered the Gold so white, might haply be rendered Red, and fix in the Gold. For every SULPHUR is a Tincture, when it is made fixed, and gets an Ingress, from the other Metals. Do not undervalue this Secret, but fish out the Property of this thing, by a more accurate Meditation, for you will draw from thence much Good.

A. Verily, I can methinks conjecture, that this very knack hath more in it than it shows for; I will search there into more accurately; perhaps this very way is a nearer one, than that which requires the inversion of the Gold. I remember that I have read amongst the Sayings of the Philosophers, this Expression; That their Gold does not tinge, unless it be first tinged, nor receives it a Red Color, unless it be made first White. I perceive, that Nature is more abundantly stored with infinite Riches, and that it cannot be so easily Searched out to the bottom, and the longer a Man seeks, the more he finds and meets with; insomuch that at last, there is such plenty of good things offering themselves to such Seekers, that it makes them puzzled which is to choose, seeing they so command each other's Benefits and Profits. Besides, your words are very hard to be understood, and hard to be born. For it seems a thing exceeding all belief, that the most constant Fire-during-Metal, Gold should be so changed, as to be no more Gold, and very hardly, yea, not at all reducible by the help of Art into its former Body. I do often meet with that Opinion and Decree of the Philosophers in my frequent reading of their Books, viz, that Gold must be putrefied, if any better and nobler thing is to be generated there-out of: But whereas it seemed unto me a thing beyond the Power of Nature, and altogether impossible, for such a constant Matter to undergo any Putrefaction, I supposed that the Philosophers pointed at some other thing by that Putrefaction of theirs. Meanwhile, I earnestly expect from you a Demonstration of the possibility and Truth of this thing.

B. Come then, on God's Name, a little nearer me, and heed well the things which shall be shown unto you.

We will here take half an Ounce of common Gold, and put it into this AQUA FORTIS, made of VITRIOL and SALTPETER, whereto we will add the same weight as the Gold is of, or a little more, of our SALT

ARMONIAC, without which, the AQUA FORTIS alone, and by itself, is not able to dissolve the Gold.

A. Pray, Sir, why do you say, OUR SAL ARMONIAC? Are there several and different kinds of it? For my part, when I dissolve Gold, I put into the AQUA FORTIS, that (common) SAL ARMONIAC, which is everywhere to be had in the Merchants Warehouses, and it is very fit to dissolve Gold into a Yellow water.

B. You speak very well after your own way; And I confess, that every SAL ARMONIAC mixt with AQUA FORTIS is very good to dissolve Gold; nor is this any new way, for 'tis in very much use amongst all the CHEMISTS, who are wont on this wise to dissolve their Gold, but yet that which is thus dissolved, still remains Gold, and doth easily admit of being again precipitated out of the AQUA FORTIS, and of being reduced by Fusion into the former Body, it had afore its Solution. But if so be, that the Solution shall be made by the help of our SAL ARMONIAC, then is the Case vastly altered, and your attempting its Reduction again will be in vain. For if Sol be but dissolved barely once with our SALT ARMONIAC, it admits not any more of melting, nor doth it of itself return again into a malleable Metallic Body, but gets a Reddish Scarlet kind of Color in the Trial (or Crucible) and remains an unfusil Powder. And if you add some BORAX thereto, and set it in the

Fire then to melt, it will pass into a Red Glass, which is a sign of its being plainly destroyed, and of its being transmuted into another Body. And therefore I dare aver, that there is seated in our SALT ARMONIAC, a power of inverting, and transmuting Gold, and of making it fit for the Philosophical putrefaction, which thing is impossible to be done by any other Salts whatever they be, and what Name soever called by.

A. Certainly, this is a Divine miraculous thing, to subject Gold, so mightily constant in the Fire, unto Putrefaction, and to reduce it by Putrefaction, into a nothing: For I have read too and again, amongst the Philosophers Writings, that it is an easier thing to make Gold by Art, than to destroy Gold made by Nature. And therefore this Salt must be a very wonderful one, which is able to effect these and other, the like almost incredible things.

B. Well may you term it a wonderful Salt, for so it is, the like of which, no Man will find in the whole World; though to such as know, it, it is so vile and mean a thing; insomuch that scarce any one would think it likely, that such things could be done thereby, as are wont to be, should it be but named by its own proper Title. Does not, I pray, that Philosopher, COSMOPOLITA (or SENDIVOGUS) confess, that he hath oftentimes declared the Art, and secret of the whole Philosophic work, word for word, sometimes to one, sometimes to another, and yet they would not at all believe him, by reason of the meanness, or vileness of the Work. And does not he make frequent mention of his own, and not the common SAL ARMONIAC? But that you may yet give more belief and credit to our Salt, I would have you read the TURBA of the Philosophers, wherein you will find all those things which they have published concerning their Salt:

And amongst others, hearken to those few words, which the Rosary mentions: OUR SALT DISSOLVES GOLD INTO A RED COLOR, AND SILVER INTO A WHITE COLOR, AND TRANSMUTES THEM OUT OF THEIR CORPOREITY INTO A SPIRITUALITY, AND WITH OUR SALT, ARE THEIR BODIES CALCINED. And for this reason, LUMEN LUMINUM, also says, THAT IF THE OMNIPOTENT GOD HAD NOT CREATED THIS SALT, THE ELIXIR COULD NOT HAVE BEEN PERFECTED, AND THE STUDY OF CHEMISTRY WOULD HAVE BEEN IN VAIN. AVICEN said, IF THOU HAST A DESIRE OF GETTING RICHES, PREPARE SALTS, THAT THEY MAY BE CHANGED INTO A CLEAR WATER, FOR BY THE FIRE ARE SALTS CHANGED INTO SPIRITS: SALTS ARE THE ROOTS OF THY WORK. HERMES said: ALL SALTS ARE ENEMIES TO OUR WORK, AND TO OUR ART, SAVE THE SALTS OF OUR LUNE: ARNOLDUS said, EVERY SALT THAT IS WELL AND RIGHTLY PREPARED, IS OF THE NATURE OF SALT ARMONIAC, AND THE WHOLE MYSTERY OF OUR ART CONSISTS IN THE PREPARATION OF

COMMON SALT: HE THEREFORE THAT KNOWS SALT, AND ITS SOLUTION, TO HIM IS THE MYSTERY OF THE ANCIENT WISE MEN KNOWN. AND THEREFORE BEND THE UTMOST MEDITATIONS OF THY WIT UPON THE NATURE OF THAT SALT ONLY, IN WHICH THE WISDOM OF THE ANCIENT WISE MEN, AND EVERY MYSTERY, IS FOUND HIDDEN AND CONCEALED. The Writings of the Philosophers are full of those and such like sayings, and they do everywhere mightily insist upon Salt. And now, what think you of these Testimonies; what do the things I have spoken, yet find any belief in your Breast?

A. Yes Verily, and now I am on your side; but yet I do as yet desire, and heartily wish for this one thing, that you would for once let me see your Labor, whereby I may convince other incredulous Persons, and make them believe too.

B. Well, I am content; and come let us go to work, and let us put the Gold in its requisite MENSTRUUM, and place it in warm Sand, thereby to hasten forward the Solution of the same; though there is strength sufficient in our MENSTRUUM, to dissolve the Gold in the Cold without Fire. We shall in a short time see it of a yellow Color: And behold that very Color, and the Gold itself is so changed, as it is never more reducible into its former golden Body. Thus have you now the entrance and beginning, which as yet is vastly distant from the wished for end: And when you now see the beginning, know, that is the first day of our Philosophic Labor. Next, let us proceed to the Putrefaction of the dissolved Gold, without which, no Colors present themselves to our view. Behold in this very moment, SOL begins to wax black, and in a little while after it will conceive such a thorough blackness, that it will be like to Ink, and may serve to write withal on Paper. This blackness; the Philosophers call the Head of the Crow, by that Name pointing out unto us their Putrefaction; by which, the second day of our Philosophical Labor is finished.

Our Ground (or Earth) therefore, being sufficiently enough moistened, we must beseech God to bestow upon us the shine of the Sun; for without the Suns heat which stirs up the Life in all things, there cannot possibly be any increase and growth. Lend me therefore your best attention. As soon as the putrefied Body of our SOL shall feel the warming heat of the Sun, its blackness, which was the true Sign of its Putrefaction, will vanish away by little and little, and give place to the access and approach of many most delicate Colors, the which, the Philosophers have named the Peacocks Tail, and this finishes the third day of our Philosophical Labor. And now, when the Fruit-producing Sun shall have thus illustrated our Field, or Ground with its warmer Rays, but for one day as yet, we may easily see, what is farther likely to come to pass hereafter.

A. Hay da! What a wonderful thing is this that I see here? In how short a time, and how speedily hath the Peacocks Tail changed itself into a thick Blood? Who could ever believe it, unless he had seen and beheld it with his own Eyes?

B. I confess it is such a thing as may well cause in any one most exceeding admiration, seeing that there proceed from Art and Nature, Operations of such great moments: God hath made all things very well, and should he not permit such an admirable changing of Color, to appear in the Operation, verily the Philosopher would be in doubt, whether or not he might hope for an happy Success of all his Labor: And now upon the occasion of this Blood-like Color, is arisen the Name of Dragons Blood amongst the Philosophers, who say, that when this Color appears in view, the fourth day of the Philosophical work or Labor is finished.

A. Good God How great and how wonderful are thy wondrous Works, who can find them out by his own Search and Industry? There do daily happen greater and greater increases of our Faith, seeing that I behold with mine Own Eyes, such things as heretofore exceeded all my belief. And I entreat you to tell me this one thing, whether or not, there lies in this Blood, as 'tis now prepared, a most excellent Power of healing Diseases or no, and if it may safely be made use of, for the healing of Sicknesses without any farther Preparation?

B. Yes verily, there does lye therein hidden, an incredible healing Faculty, and this, its very Signature doth openly witness. For God doth not sign or mark all the Herbs, all Animals, all Stones in vain, by the external Signature of all which, it is an easy matter to know, what use they are of for Mankind. Forasmuch therefore, as our Gold is turned into a thick Blood, it is a most certain Testimony, that it is Serviceable for the curing and amending of Man's Blood. For, if the most ancient Physicians have made use of Gold, reduced either by filing into a most subtle Powder, or by beating into most thin Leaves, in the Cure of the most desperate Diseases, yea, and Of the Leprosy itself too, and this not without Fruit, and Success; how much greater commerce then will this Gold of ours thus changed into a Red Blood, have with the Blood of Mankind. Would it not (think you) abound with a greater Power of freeing it from all false and superfluous Aqueity, and other evil Humors and Impurities, and of reducing it into its former State of Good Health? For the occult Virtues of the Gold are no more shut up, now, like as when the Corporeal filed or leaf Gold held them locked up, that they are unloosened, and released from their Fetters, and so perform their work by most free and unclogged Operations. And seeing that Gold is, by Ancient and Modern Physicians, attributed to the Heart, and to the whole blood of Man's Body, and seeing that experience itself hath demonstrated, that it takes away the preternatural effects of the Heart, and heals the corrupted blood; why should not this our Blood of Gold effect the same with greater Efficacy, and a more happy Success. With this very blood, have I already performed wonderful things in the most grievous Diseases in the Bodies of Men, after they have been first purged with some PANACEA. I have therewithal healed the Leprous in a short space of time, and such as have to no purpose made use of the assistance of the most eminent Physicians for ten or twelve years continually, to their great Costs and Charges, and this I can evidence by most clear Testimonies, which I keep by me, Nay farther, even at this very day in very many places, into which I have sent of this Blood, the detestable Leprosy, the FRENCH POX, and other most grievous Diseases of the Heart and Blood (to the Cure whereof, Gold is in an especial manner ordained by God) are taken away by an admirable way of healing: For seeing, that (as we said afore) it strengthens the Heart, and mundifies the defilement of corrupt and polluted Blood, and purges them away, it must of necessity likewise expel, as well the open or visible, as the occult and hidden Diseases of the Body of Man. For if Nature be but corroborated, it doth easily subdue and expel Diseases, by the help of other very small Medicaments; the which thing I have by my manifold experiences found to be most true, when I have adjoined this Aureous Blood to other suitable Medicaments. For it cannot be expressed, what, and how much I have done, both in young and old, with this very Blood so exhibited, in the STONE, GOUT, DROPSY, EPILEPSY, and other Chronic Diseases, which have taken deep Rooting in the Body. Therefore if God shall please to lengthen out my days, I will very suddenly publish the use thereof, that so it may be brought into common Use in all Diseases. For this Golden Blood may (probably) be the potable Gold of the Ancients, which never more suffers itself to be reduced into its former malleable Body. I have sundry ways attempted to reduce it, but never could effect the same. But yet this one Case I except, viz, a little of this dry Blood, being put upon molten Gold, hath ingress into the same, and the residue swims at the top thereof like in Earth: but yet that little which adjoined itself to the Gold, is of so great a Power as to make all that whole Body of Gold which it entered into, brittle, yea so brittle, as that it suffers itself to be beaten in a Mortar into most fine Powder.

A. I Marry, Sir These are Miracles indeed, which I see and hear; who will make any farther doubt, but that the Universal Tincture which heals the Sicknesses, or Distempers both of Men and Metals, may be prepared, out of this Golden Blood? For all the Philosophers do with one Consent confess that their Tincture, when quite perfected and cast in upon molten Gold, doth render the same brittle. And now seeing this Golden Blood of ours, being as yet immature, and not prepared, doth effect the same, would it not, I pray, perform the same much better, if it had but ingress given unto it, by inceration, whereby it might flow the easier, and enter the more readily. I do not now at all doubt, but that this aureous Blood both can and in time will become a universal Tincture full of Medicinal Virtues.

B. Although I am not minded publicly to disclose an ARCANUM of such great moment, and so great a Mystery, and to throw such a precious Pearl afore Swine; yet I am of the mind to prepare some quantity thereof, and to part with so much unto the Sick that need it, and that shall desire the same from me, as is requisite for their Use. And not only to the Sick, but to others too, that are willing to apply it to other Uses and Experiments; but especially unto those who would fain try, whether or not, even this very Blood will turn, itself (according to my Description) into a white Milk, and then into a Red Stone, and, by a new Reiteration of the work, pass through all Colors. This desirousness or inquiry after the Truth cannot at all prove any ways disadvantageous to me, or mine. For the Art itself will always remain an ART, unless it chance to happen, that someone or other Searcher, which is of a more subtitle Ingenuity, should, by his diligent and serious inquiry, search out the very Foundation of the Art itself. Which if it should so come to pass, he must then think, that God hath vouchsafed this Gift unto him, and that he now considers, and well knows, what esteem he ought to set upon it, and by what means he ought to hide it. However, this is most certain, and indubitably true, that this Doctrine, and instruction of mine own prescribed in my Writings, will be a spurring encouragement to abundance of Men (that have hitherto exceedingly doubted of the Truth of this so great a Mystery, and shall now come to know it, laid open by me with such evident Demonstrations) and cause them to search with a greater and more serious study, and to cast off all other unprofitable Coctions, whatsoever. For to what purpose is it to use many things, about that which may be done by fewer and more easy. All things have their time, and so has this ARCANUM too its own proper time. But we shall speak more of these things in the following

Description of the second universal Medicine. All these things which thou hast hitherto seen are indeed very good; but we are as yet far off from the end. Have you not likewise read in the Philosophers Writings, that the white Swan doth also show itself in the Work? Now, if by but as yet ONE days shining, the Sun shall have illuminated by its brightness the Dragons Blood, you shall see it turned into a white Milk, which Milk the Philosophers have written off, and which at length goes into a COAGULUM, or Cheese like Curdling. Look therefore now upon that Milk, which you see to admit of Coagulation, and Condension, by little and little: And thus with this golden Cheese do we finish the Philosophical Labor of the fifth day.

A. God be thanked, that this days Labor hath also succeeded, as we could wish. But forasmuch as those things which you mentioned, but now come in my mind, viz, that we are far off from the end of the work as yet, and yet tomorrow is the Sixth day of this our Philosophical week; and farther, seeing you said presently after the beginning of this our Conference, that the whole Work would be finished in six Days space, and that on the seventh Day we may cease from all our Work and Labors, and sanctify it, or keep it Holy, and give God due thanks for all these his Benefits bestowed on US: Seeing (I say) that all these things come now to my mind, 'tis no wonder, if they likewise create in me a great deal of Care and Puzzling, to think how this can be, that all these things that remain yet behind may be perfected, and brought to a full end in one day.

B. Cease your Care (my Friend) nor do you ought else this Sixth Day, but hourly increase the Fire by Degrees, and stir it up more and more, that so you may see, by what means our white COAGULATED Milk will by little and little pass into a yellow Color, and will at length be thoroughly Red, and abide most constant in the Fire. This fixed Redness, the Philosophers call their SALAMANDER: The Poets tells us a Tale, of a certain Worm that lives in the Fire, which cannot be burnt or consumed thereby. Therefore, after the Philosophers have brought their work through all the Colors, and have at length attained so far, as that there appears no other, but a mere fixed Redness, they named it their SALAMANDER, with which (if you except only Inceration and Multiplication) they ended their work, and so do we also finish our present Labors.

A. Praise and Glory be to God, by whose Divine help, we have (by so happy and desired a Success) promoted our work to its most desired end.

B. AMEN. And thus will we now put an end to our work, that so we may sanctify tomorrow, which is the seventh Day, to the Honor and Glory of the Divine Bounty.

A. Ah, my dearest Friend; let me entreat you
not to involve me by your immature departure, in
greater troubles and difficulties. There are divers
scruples and doubts, and those weighty ones too,
that perplex my mind, which unless you remove afore
you're going away (but I hope you'll stay) I shall
of a certain truth be tormented all this ensuing
Night with the most bitter Pill of Disquiet and
Anxiety, and then you may well guess with what mind
I am likely to celebrate the tomorrow Sabbath. For I
am yet as plainly ignorant, what use to put that
SALAMANDER to. As concerning those things, which you
have faithfully disclosed unto me hitherto, I trust,
I shall not err in their Operation, but as touching
Inceration and Multiplication, in which, as in two
Cardinal main Points, the very (Pillar or) hinge of
the whole Operation yes, as you said, I must needs
confess myself more Blind, as to them, than TIRESIAS
was: I must needs say, that I behold the promised
Land situate afore mine Eyes, but the way that leads
thereunto is bridged in with such Thickets, and so
many Brambles, that I do not see which way I shall
extricate myself out of them. Unless the mercy of
God, and your help come in to my assistance, I see
that all my Labors will be in vain.

B. 'Tis no small trouble you bring me, by your
importunity, don't you see the approaching Evening.
You act just according to the Custom of importunate

146

Men, who having once gotten ones out-held Finger, do snatch in the whole hand. At first, you only entreated me to discover unto you the Matter and Key of the Art, and said that you would easily find out the rest yourself.

Why then do you not seek thereafter, and let me go?

A. Good Sir, be not displeased with this my importunity, proceeding from the too earnest desire, I have, of knowing so great a Secret: And Christ himself said, if ye shall knock, the Gate shall be opened unto you.

B. Well, since I see, that I must expect no quiet from you, till you do likewise learn something from me, concerning Inceration and Multiplication, I will in a few words set afore your Eyes, things of great moment. Look to it, that you listen very attentively.

A. I do.

B. Have you not read in the Philosophers, when they speak of Inceration, that the out-driven Soul is to be restored to the dead King, that the dead Body may be recalled back to life, and that it, arising with a more glorious Body, and a more excellent Crown, may prove a helper to its meaner Brethren. The Philosophers words are, as follows. Here the Soul lets itself down, and refreshes the dead Body. For it is not sufficient, that the King be deprived of Life and so left dead: No, no, for necessity requires, that its Soul be restored unto it, which may restore its Motion, and lost Life, to the dead Body. Now, by how much the oftener, the Soul and Life is taken away from the King, and that which is taken away be again restored thereunto, which so much the stronger and more active Body, and so much the Magnificence a Crown will he arise withal. By these few words have I laid open unto you, Inceration and Multiplication. But yet there are other ways of increasing our fixt SALAMANDER, and rendering it fusible, viz, by the addition of Mercurial things, which, by their speedy Flux and penetrating Property, do pierce into this our destroyed Gold, dissolve it, and so bring to pass, that there is made of them both (viz, of the destroyed Gold, and which admits not of any reduction, and of the Volatile MERCURY) a certain fusile middling Body, which said Body, thus conjoined of the two, is to be maturated by the bare Regiment of the Fire. And by this Maturation, is this universal Medicament rendered so fusible, as to have Ingress into all the Metals, and to penetrate them.

A. (But pray Sir,) Is not this way of giving a more easy Ingress and Flux to our destroyed, and irreducible, Gold, by the MERCURY of Metals, more facile, and a nearer one than that above said way, which requires a great many Operations, by the reiterating of Inceration and Multiplication?

B. Yes Verily, it is a shorter and easier way, as being void of many tedious Labors, for it needs nothing else, but that, the MERCURY of some Metals be put into some good strong Glass with the inverted Gold, and be so brought into Fixation. But yet this Medicament, that is on this wise wrought up with the MERCURY to a constancy in the Fire, cannot extend its Color so largely, as that, which is rendered fusible by so many reiterated Operations, because, in every reiteration, the Tincture is exalted and multiplied. And now, have you any other demands? If so, be brief, for the Evening approaches.

A. Yes, Sir, there are many things that I would ask about but seeing that my importunity is troublesome, I will at present rest content with those things, I have heard. Only, this one thing more would I gladly know, viz, where I ought to seek for the Soul of the King.

B. You must look, whither you have driven it, and there must you seek it, and, having found it, you must restore it to the dead King, and so you will again begin your Work, and you shall again bring it through all the Variety of Colors, like as you did at the first time. For when the Soul is restored to the Body, there is made a new Solution, which is to be again putrefied, that it may turn black; and then proceeding farther on according to the same way, as was done in the first Operation, there will appear all the Colors, and they too far more delicate than in the foregoing Labor. The CROWS HEAD will be Blacker, the DRAGONS BLOOD Redder: The LAC VIRGINIS Whiter, and the SALAMANDER more fusile, than it was in the first Operation. For by how much the more often you shall repeat this Mortification and Vivification of the King, so much the more Magnificent, more precious, and more efficacious a Tincture, will you obtain. Believe it, and give God the thanks, and be mindful of the Poor, as soon as ever you are Master of your Desires. Come to me again upon MONDAY, that I may also teach you (according to my Promise) the way of making another Philosophical work out of the Poisonous, Volatile, and black SATURNINE MAGNESIA: And so farewell.

A. In going away spoke thus with himself, Praise and Glory be unto God, who hath thus, by the help of one Man, bestowed on me so great a Grace. I now go home with a cheerful mind, and there will I most desirously wait for MONDAYS approach, that so I may likewise get, from my Friend sent me by God, that other work, (made) out of the Volatile and Philosophical Gold. But this Operation, done with fixed Gold, doth please me wonderfully well, especially for this reason, because, in every Reiteration, there do always appear the same Colors, which presented themselves to view in the first Operation. Who would ever have believed, that so many, and such excellent Colors lay hid in common Gold? And now, if this can be demonstrated in the common Gold, what may be expected from the secret Gold of the Philosophers, in which (they say) many more Colors lye hid than do in common Gold. 'Tis a common Proverb, (I confess) that he who seeks from another comes to know many things, but (usually) renders himself ungrateful: But yet however, I cannot choose but propose unto the Man one Question more, perhaps some sweet refreshing Solar Ray will shine upon me. My Friends I am come again unto you.

B. Well, tell me, what more you have to say unto me?

A. As touching those things which I have seen, I desire no further Information concerning them. But, it will most thoroughly serve my turn, if you will be but pleased to answer but unto me one Question, which I shall propose unto you: And that is this, Forasmuch as almost all the Philosophers, in the description of their Stone, do tell us, That the King is to be conjoined, and Radically mixed, with the Queen in a peculiar Mercurial Bath, that so from them an Off-spring more noble than the Father may arise: And seeing likewise, that the same Philosophers do at large celebrate the Coition of MARS and VENUS:

My desire is very great, to understand your Opinion concerning this business. For if there could happen, or proceed, any good effect from this Conjunction, of Gold with Silver, or MARS with VENUS, what need would there be of so many repeated Coctions of the Gold?

B. This way, which you make mention of, is not unprofitable; and, to tell you the Truth, there are yet nearer ways, in the which whosoever enters into, will equally become a Master of great ARCANA'S. I was willing to show unto you the work upon bare Gold only, for this reason, viz, that you may see with your Eyes, and as it were feel with your hands, that those most eminent Colors do lye hid even in bare Gold too; which is a thing, that scarce one of a hundred or a thousand would have believed. And now being certainly assured, about the lying hid of so great a Tincture in Gold, you may with the greater confidence set about your Operations.

A. Still more and more doth a brighter light shine upon me, Nor can I enough admire, when I look upon the Rosy Color of the Lion, that out of so small a Lion, there should proceed so great a quantity of Blood, colored with so excellent and delicate a Color, when as notwithstanding, out of other much greater Animals, there is scarce extractible so much Blood, as to answer to the hundredth part of this Blood.

B. What? Do not you understand, by what means, such an incredible quantity of Blood can flow forth out of so small a Lion? What, do you not know, that the whole Body of the Lion, which is the King of all Beasts, is nothing else but the mere concentrated Blood of the heart, one half Ounce of which, being boiled in many hundred Ounces of water, doth turn all that water into Blood, as you yourself have now seen with your own Eyes.

A. I confess, that there have now been shown to the view of mine own Eyes, such and so great things, the like of which I never heard so much as one word of, from any others. No Body would believe me, and (which is worse) mine Enemies, who are most vile Compilers of infamous Libels, would set upon me with store of reproaches and lies, and call me Sophister and Cheater, and would say, that I boil some BRASILE wood, or other Red coloring Stuff in water, and so sell it for the Rosy Blood of the Lion. And therefore, I have no reason to make show of any knowledge of this so great a Secret.

B. How? What will you be afraid to be a reporter of the Truth; and to manifest these great wonders of God and Nature. That very way by which I showed unto you, that that most high Color arises out of Gold, by the self-same way may you be able to shut the defiled mouths of your Enemies. And if they will not believe, that there lies hidden such a

Color in Gold, they will be convinced by the said Lions Blood, which, being made bright hot in a strong Fire, and afterwards boiled again in other water, will convert the same, as it did the former, into Blood, and this you may even do, as often as you please. By this it is manifestly evident, that the Blood of our Lion is incombustible. And in our being busied about those Labors of ours, there present themselves to our view, three most delicate Flowers: The first is, a VIOLET, dyed with a mixt red and Sky Color: Then, the White LILLY; and at last, the immortal AMARANTHUS shining with a Scarlet dye. But why, immortal?

For this reason, because neither the strongest Fire, nor the strongest Corrosive waters can at all hurt it, or change its Colors. The AMARANTHUS is tinged with the highest and most constant Color, beyond all other Herbs and Flowers, which does not so soon vanish after the manner of other Colors, which other Flowers have, but abides constant for many years, insomuch, that such an Herb is kept both Winter and Summer, to make Garland and Posies, or such like, withal.

A. If now anyone could be so happy at last, as to enjoy that AMARANTHUS, he might present many pious Virgins, with Posies and Garlands made thereof.

B. If you are desirous of using the Lions Blood in your Operations, then follow PARACELSUS his Doctrine, in his TINCTURE OF THE PHILOSOPHERS, where he teaches, that the Rosy Blood is to be conjoined and fixed with the white Glue of the Eagle; which way seems the nearest for the obtaining of the universal Tincture.

A. I am delighted with hearing of this: The Rosy Blood of the Lion I have, but yet I want the white Glue of the Eagle; which if I could but get into my power, I would conjoin them both together without any more delay, and bring them to Fixation. But I cannot sufficiently enough admire at this, viz, that PARACELSUS hath made no mention, of conjoining the Virgins Milk of DIANA with the Kings Blood.

B. PARACELSUS has not therefore injured any, in his not manifesting all things so clearly and openly, seeing it is the Custom with other Philosophers to do the like: If you are desirous to use Gold and Silver conjoined together, in your working, then deal with the Silver, just as I taught you to do with the Gold, and make thereof an incombustible LAC VIRGINIS, and such as admits not of any Reduction; the which is to be conjoined with the incombustible Blood of the Lion, and to be put into a Vessel, that so these two Bodies, being conjoined, may never be separated from each other by a kind of disjoining. But, that you may yet better understand the Whole business, I will subjoin a few words more, Our White Eagle being put in common Water, makes the same, Whereas it is dissolved so tenacious and glutinous as that, it can glue Paper or Other things together; and therefore it is not Without cause, that this Our secret Salt is termed by Philosophers, the Glue of the Eagle, For those things, which this Glue Joins together can never more be separated from each other, and this is one mutual Bond, by Which the Husband and the Wife are so linked together that they cannot separate from one another. It doth not much differ from the tie of Marriage, by Which, the Husband and the Wife, or the young Man and Maid, are by Gods Command so knit and bound to each Other by the Ordinary Minister of God, that nothing but Death can separate them. If now a Man and Woman join themselves together without the legitimate and public Bond of Matrimony, they beget illegitimate Children, and can (When they Please) break that Conjunction of theirs and depart the one from the Other, because they are not Coupled with that Marriage that proceeds from the Command and Will of God; which said Coupling or Matrimony is in repute and is generally observed for a common Custom and as being the Will and Command of God, not only amongst Us CHRISTIANS but likewise amongst JEWS, TURKS, and HEATHENS, throughout the Whole World. And for this reason, the Philosophers were Pleased to introduce the like Coupling or Marriage in their Work, and that not only of SOL with LUNA, but likewise of MARS with VENUS, Which Copulation if wanting in these two last (viz, MARS and VENUS) their Conjoining cannot have any good and happy Success as daily experience witnesses. But flow if MARS and VENUS be Coupled by a Priest in the lawful Bond of Matrimony they likewise bring forth legitimate Children and such as may be Promoted to Kingly Dignities I could yet kindle more light unto you here, concerning this Our Philosophical Copulation viz, by What means the most inconstant and most highly Volatile things may (by the operation of the same) be so conjoined, as that they can never afterwards be loosened from their inseparable knitting together. But at the present you are not capable of bearing any more. Whatsoever hath been now omitted, by reason of the shortness of time, shall be made up at our next meeting and Conference. But, one thing more comes to mind, which I would have you know; and 'tis this. When you would make any good thing of MARS and VENUS, you must in an especial manner beware of their gross Bodies, which are Leprous and unfit for our work. 'Tis their Blood you must seek after, the Blood, I say, of their inmost Heart, and the true Soul that lies hidden in that Blood, which is so very near a kin to the Royal Blood; And now, that you may enjoy such a most noble Blood, you must not imitate the Country fellows, who, thrusting a knife into the Hogs necks, do save that Blood that flows out from thence, to make Puddings withal, and take both good and bad Blood together. No! You must therefore in the first place, know for a certain Truth, that you must separate out of that grosser Blood of MARS and VENUS, that most subtle Arterial Blood of the Heart, without which separation, you may not expect any good succeeding Event. Which said Separation, seeing, I know, that you are ignorant of, I have thought good to set it afore your Eyes, which more clear and more perspicuous Expressions. (Know then) that that most pure Blood is never gotten by any other means, but by the Corruption and Destruction of the Veneral and Martial Body: By which helps, these, so gross and hard Bodies are in such wise opened, that the inmost and most subtle blood may be drawn out of them. But yet, there's no need of destroying their Bodies by AQUA FORTIS, seeing that Nature herself exhibits unto us their Bodies already opened, in VITRIOL, and hath left no more for us to do, but the Labor of extracting the most pure part there out of. For even in the common VITRIOL, MARS and VENUS are found conjoined by Nature. And now, when we would draw there out of the most pure Blood, such a VITRIOL is to be chosen, which never hath been as yet in the Fire; this, VITRIOL is to be dissolved in water, and to be filtered and separated from all its Feces. This green VITRIOL contains hidden in its self, the Blood of that green Lion, out of which, the said blood is to be extracted by the help of a certain Magnet, that so the unprofitable and gross Body may remain behind, For 'tis the Spirit that vivifies, the Body is unprofitable. But now, in your extracting this Blood thus pure, you must not be too greedy, but have a Care, that you extract no more than a little of the best Blood: Which if you do not heedfully mind, you will also extract together with the pure Blood, the more gross and more impure, and so your work will be spoiled. For you need only the best and most subtle white Blood of MARS and VENUS. For, like as the subtlest and noblest Blood of all the Animals is not Red, but White, (viz, the Sperm): Even so, the Blood of MARS and VENUS, which comprehends the Virtues of the whole Body, and the true propagating Seed, is tincted with a white Color, in the search of which, 'tis expedient we bestow our Labor. Therefore, after that we have extracted, by the help of our Magnet, some two or three half Ounces of white Blood, out of twenty Pounds of VITRIOL dissolved in water, we then take it out, lest it should also extract the most gross Blood too, when it can find no more of the white Blood unextracted. You had therefore need be cautious in the extraction of that Tincture, and be circumspect in the Accomplishment of the same. For we are taught by Nature and Experience, that, in the Coition of Male and Female, there doth naturally concur a white Blood to the Propagation of their Off-spring, and in defect of that, Nature is constrained to send forth a vile and red Blood. Therefore, we must here endeavor, with our utmost Care and Industry, to purchase by our extraction the most pure Whiteness, and not the most gross Redness. For in that white Virgins Milk is hidden all Colors, and the highest Redness. This white Virgins Milk, you may promote to a Fixity and Constancy, in a pure clean Glass, without the Addition of any other thing, the which, after its being dried, will become black, and will be translated out of the black Color into several others, and at length shines with the highest Redness, which puts an end to the Operation.

But as concerning our Magnet, seeing you know it already, you will not need any farther instruction about it. You have thus heard my Opinion about MARS and VENUS, and have understood, by what means some good thing may be thence had. Neither must you persuade yourself, that there is any other way of having any benefit out of them; how great is the number of those, who by their vain Labors have mightily damaged themselves, following the Writings of BASILIUS, in which he teaches to make a Red Oil out of VITRIOL, which he calls Gold Philosophical. The bare literal Sense is not always wise, and everywhere in his Writings care needs to be taken. There is no doubt, but that no small benefits are to be had out of VITRIOL, For the Philosophers themselves do hint forth unto us, the benefit therein hidden, in these words: VISITABIS, INTERIORA, TERRA, RECTIFICANDO, INVENIES, OCCULTUM, LAPIDEIVI, VERAM, MEDICINAM; YOU SHALL VISIT THE INWARD PARTS OF THE EARTH, AND BY RECTIFICATION YOU SHALL FIND AN HIDDEN STONE, A TRUE MEDICINE. And now at last, to close up this our Conference with some profitable Admonition, know, that I would have thee forbear from the troublesome Coction of the imperfect Metals, by reason of the too many impurities, which create many Molestations in the Separation of them, and cause much loss of time. 'Tis better, that you take in hand SOL and LUNA to perfect the Work withal, which Metals do most abundantly contain in them, the Red and White Tincture, and are already freed by Nature from most of their impurities. And although, I have not as yet made the universal Tincture out of SOL and LUNA, yet there hath appeared unto me so much from so many and such various Operations, that I do not in the least doubt, but that the immortal AMARANTHUS may be made out of them; whereunto you may boldly give Credence, without doubting it at all.

The COROLLARY, Or, Present over and above.

In this Dialogue, the Dragons Blood is so described, as if it needed full six whole days Operations, to be perfected and brought to its full end, nor did I then know any nearer way. But afterwards, there became known unto me a nearer and more compendious Process, whereby I can now (Praise and Glory be unto God for the same) in a few hours' time introvert and transmute the dissolved Gold, and reduce it into a Red Blood, and that by the help of one only certain White URINOUS sublimate, concerning which, I have treated more at large in the fifth Part of my PHARMACOPEIA SPAGYRICA. And upon the same Account have I likewise written, that the said Blood of Gold must (needs) at length get an easy Fluxibility and Ingress by the frequent Iteration, or Abstraction of the moist Spirits from the dry Gold. But forasmuch as the Multiplicity of such Abstractions is not only full of laborious Operations, but also chargeable and requires much time; and that the studious Searcher after so great a Secret, may not by being affected with the Tediousness of so many toilsome Operations, be quite weary, and at last throw off all hopes:

I do therefore commend unto every one, that other and nearer way of getting it a Fusibility and Ingress, viz, the incerating it by MERCURY, by the benefit whereof, he will far sooner and easier arrive unto his purpose and final Scope. And although I have hitherto been hindered through want of time, besides (many other impediments) from having completely finished the Operation of the said Tincture. Yet notwithstanding, I do with an assured and undoubted confidence affirm unto every one, that there is given a most certain occasion (or Capability) of the making and finishing it, and that by the hitherto described way. I bequeath to such of our Posterity, as are of a Good, Sincere, and faithful mind, those things which I have wrought with mine own hands and withal to what issue I have brought all my Operations. It is free for every one (as time and occasion serves) to make Trial about attaining unto the wished for end. For my part, I do heartily thank God, that he hath vouchsafed unto me the Ability of plentifully and abundantly preparing the Lions Blood, and of performing no small matters therewith in a Physical way:

And therefore I shall be well content, though I never arrive unto the desired end of finishing the Operation of the universal Tincture.

The Second Dialogue

The Second Dialogue, or Conference, Or a Continuation of the precedent Colloquy or Dialogue, treating of the Preparation of a universal Medicament out of the black venomous and Volatile, SATURNINE MAGNESIA.

A. Blessed be the name of the Lord, who hath caused another Day to shine upon me. I will presently haste to my faithful Master, and will beg from him the Doctrine of the second universal Medicament, that he promised me.

All hail thee, my dearest of Friends! God Grant that this Day may prove happy and lucky to me and you: I am come hither to hear from you, if it stand with your Convenience, to make good your Courteous Promises at this present, and to teach me the making and Preparation of your second universal Medicament:

For I have an exceeding ardent desire, to know the manner of Preparing it, out of that Volatile and poisonous Mineral.

B. I thank you for your good Wishes, and wish unto you the same you wished me. Look here, here's a piece of our venomous SATURNINE MAGNESIA, which is the true Matter that the Philosophers Stone is prepared out of.

A. I pray, is this black stone the true Matter, out of which is wont to be made the universal Tincture, and Stone of the wise Men? Verily I much wonder, that this should be the Gold of the Philosophers, I am extraordinarily desirous of seeing, by which means so excellent a Medicament, and so noble a Tincture can arise out of so base and venomous a Body. I should rather adhere to that old and common Proverb, and say, who can wash the BLACKMOORE white, which Nature hath generated Black? This now seems unto me more estranged from the Truth, then that which you propounded last week of the common Gold. For how could it seem at all likely to any ones bare reason, that such a Body could be brought into a nothing, and again reduced into a certain Body: But time will instruct, whether this be possible or not. I will therefore very patiently wait for the Event of those things, which you shall show unto me.

B. What? Still more incredulity, and do you anew produce (and Practice) the Faith of unbelieving THOMAS (viz, to see and feel afore you believe?). What do you think, I would go about to persuade you, that you have a wooden Nose sticking on to your Face? Alas, Sire My time in a little more precious than so, to spend it in an unprofitable Tattling. And that time which I now bestow upon this our Conference, is spent to this end, to free you from your unbelief, and to bring you unto a true and evident Credence; to the performance of which, I am stirred up by the Authority of the Sacred Writ, which Commands, that we reclaim such as err, into the right way, and that we shall receive from God this reward for our Labor, viz, our Star shall shine in Heaven brighter than others. Besides, Christian love requires the same at our hands, that we do good to others. And forasmuch as I have well known your Christian like Conversation, and Godly way of living these many years past, why should I deny you the things you demand, or wind you into the Intricacies of a greater Labyrinth, seeing you have but too long stuck in them already? The things which you cannot at first understand or believe, the end will at last constrain you thereunto. Your part is to listen very attentively to all the words, I am about to speak. For I can easily take away from you all incredulity, and remove out of your mind every Scruple of doubting. How sourly you look upon this black Mineral, well, but you shall presently find, that all the most delicate Colors of the whole World are most abundantly found therein; and by the help of an Art described by the Philosophers, they will appear one after the other in Operation, even from the Black head of the CROW, to the Red SALAMANDER.

A. Bless me, what an ardent desire have I to see these things?

B. Have you not read in the Philosophers, that the pure is to be separated from the impure, and the purer part is to be ripened? Separate, say they, the pure from the impure, and bring it to Maturity. And they call Separation, the washing away of the Blackness, which being washed off, the Whiteness presents itself to view? The Fire (say they) and AZOTH wash LATON; LATON signifies our black Mineral: AZOTH signifies their strong ACETUM, which wets not the hands. This ACETUM, as well as our SAL ARMONIAC (aforementioned) hath its Rise from common Salt. Both of them, as to the external shape, and taste too, and Efficacy and Virtue are in a manner just alike. With such an ACETUM is our black Matter dissolved into a Snowy—white, pellucid and very clear, water, as you may here see. This water have the Philosophers called their MERCURY: In it, are hidden all the Colors that are in the whole World, but yet not visibly evident, afore that this water, or this MERCURY be put upon the water, (I believe he means, upon the Fire) and be cherished by little and little, like Hens Eggs, which cherishing dries up the water by little and little, into Earth, and renders the Colors visible one after another, as you shall presently see.

A. (Good luck) who would ever have believed, the things which I at present see? Our Mercurial water hath been scarce three Days in digestion, yet begins to change itself into a White Milk. And now it is (but) the eighth day; and this said Milk goes into a COAGULUM or Curd, and within in the Glass about the Edges (of the Matter) there shines a delicate Redness, but yet I believe it is fixed or Constant.

B. Pray, how can it be constant, seeing it does but represent the DRAGONS BLOOD and will presently be gone? But proceed you but on with this first Degree of Fire only, and that little and little, until the whole water be turned into an ashy Colored earth: When this is done, we will increase the Fire by little and little, one Degree more, which will leisurely turn the ashy Color into a Black one.

A. I will use my utmost diligence and observance.

B. Do you see now that sleek and shining Blackness like the Head of the CROW, covered over with abundance of black and very small Feathers: And upon this Account, the Philosophers have called this thus appearing Color, the CROWS HEAD. To this black CROWS HEAD, administer the third Degree of Fire, which will transmute this black head into various, most delicate Colors, shining like SOL and LUNA: Then continue on this degree of Fire, that all the said unstable Colors may vanish, and may present to view the white Color. After Whiteness, follows Yellowness, which at last will be covered over with the constant and permanent Redness: Which appearing, the fourth Degree of Fire is to be administered, that that Redness may be more and more exalted, and waxing Redder may attain unto its due Fixity and Constancy; the which by way of similitude, the Philosophers call, a SALAMANDER, and is the end of our whole Work.

A. I should never have so much as dreamed, that so black a Body could in so short a space of time have been transmuted into a most pure Whiteness, and that this same Whiteness, could, by an admirable passing through all Colors, pass into a most delicate Redness, but I pray, when this Matter hath obtained this Redness, is it to be accepted of, as an Universal Medicine?

B. Yes, Verily, because all the curable Diseases of Men, may (by that same Matter) be restored to former Health; and that safer, better and more perfect than can be done with any Herbs, or any other known Medicaments. I except that Medicament, which is extracted out of this, and concentrated into a much nobler Nature. But, as concerning the Transmutation of Metals, it yields not any Profit, afore it be made fixt and constant in the Fire; to the effecting of which, there is required a sufficiently great space of time.

A. As far as I can understand, there may be a yet better and more profitable Medicament prepared, than this is.

B. Yes, Verily, that there may, a much better and more useful, because that with this, there are as yet admixed many unprofitable Feces, which ought to be separated therefrom: And the Case is far otherwise in the perfecting of this Work out of this Mineral, then in that which is done with Gold, for this is all over defiled with many impurities: For everyone must needs think, that there are abundance of Feces (that are to be) separated from that Mineral, and by how much the more impurities are separated, so much the more efficacious must the Medicament itself necessity be. Now in this present Degree and State it is brought unto us, it would be sufficient for all kind of Diseases; which if we would yet have to be far more efficacious and stronger, it would be expedient, that we separate yet the more unprofitable and useless Feces, and concentrate the more pure Essence into a more narrow Room and lesser Body. For 'tis the Soul only or Quintessence of things, that heals Diseases. The shells or husks bring no Profit at all, and this the Husbandmen well know; for they separate the husks and chaff from the Corn, afore they bake Bread. The Medicinal Virtues of Herbs and Minerals are but of a small weight, afore they are freed from the Bonds

and Fetters of their Bodies: But now after that they
are separated from their Bodies, they can perform
incredible effects even in a most small quantity,
and such as the great weight of that Body, whence
they are extracted, will never effect. Look but on a
living Man that is in good Health, with how ready
and nimble a Motion can he stir his Limbs, and what
strength he can put forth: But as soon as ever the
Soul shall have separated itself from the Body, how
insensible and immoveable the Body lies, and not
serviceable for any Use? It is therefore a certain
and undoubted Truth, that the Life of all things
want weight, and this shall be more clearly and more
evidently demonstrated by the Concentration of this
universal Medicament. For that which one Ounce, now,
does, of this thus prepared Medicament; half an
Ounce, when concentrated, will perform the same: And
that which half an Ounce of this Medicament once
concentrated will effect; one quarter part of an
Ounce, or a Dram of the same twice concentrated,
will effect the same. And according to this compute,
may you proceed on farther. For by how much the more
often the prepared Medicament is concentrated, so
much the more Feces are separated therefrom: And by
how much narrower the Virtues are contracted, so
much the greater effects do they produce. And thus
there needs not in a manner any weight in Medicinal
use. That which ten Grains of an unconcentrated

Medicament is wont to accomplish, you will effect the same with one Grain of the Medicine, when concentrated, to be put either in Ale or Wine for some hours, if need be, or else held in the Mouth. For so it will no less display its occult Virtues and Powers, than if the Powder of the not concentrated Medicament had been otherwise drunk down. Besides too, such a Medicament may be a long time used without losing of its Virtues, and that not only inwardly, but outwardly also in all Wounds, Ulcers, and such like external effects. For all new Wounds, as also old Ulcers whatsoever they are, are happily cured by the laying on of those Medicaments, if the same Medicines be likewise used inwardly. Nor needs there be many EMPLASTERS, CATAPLASNES, and Ointments: Concerning which things, see more in the Fifth Part of my PHARMACOPEIA SPAGYRICA, where the use of this Medicament is treated of.

Read also those things, which that most excellent Philosopher VAN HELMONT declares of another certain Philosopher, BUTLER by Name, viz, that he had a certain Stone, at LONDON, in ENGLAND, which being swelled a very little in Oil Olive, rendered the same exceedingly Virtuous, that some few drops thereof taken into the Body, would drive away the most grievous Diseases, and being outwardly applied to Wounds (or Sores) would speedily heal them. These things HELMONT testifies to be true, as

172

being an Eye witness of the same. But I do not
attribute such incredible Virtues unto my
Medicament, for as much as I myself doubted of this
Story of BUTLER, and could not believe, that any
Medicament could be promoted to this so high a
Degree of powerful Efficacy: But now, seeing I
perceive that it is possible to Art, for the Virtues
of things to be contracted into a narrow Room, and
be concentrated; I do upon good ground cast this
doubting off from my mind, and adhere to those
things which I see with mine Eyes, and touch with my
Hands. Certain, and firmly undoubted it therefore
is, that not only the Medicinal Virtues and Powers
of this our Matter may be contracted into such a
narrow Compass, as to effect an hundred fold more in
Medicine, than that more gross Body could do: But
also this may be acquired or effected too, viz, the
hidden Color or Tincture in our black MAGNESIA, may
be likewise be concentrated, together with that
Multiplication of Virtues, insomuch, that in the
several Concentrations, the Color of the
concentrated Body may be exalted, the which, most
high Color, or exalted Tincture can never at any
time be gotten, without our secret Concentration:
For other-wise, if there were no need of that
Concentration, it would necessarily follow, that
that MAGNESIA of ours would be no other than an most
pure Body, not at all needing any ablution of its

gross, and unprofitable, and black impurity: But this is not so, as we said before about the Gold, which is pure in its own Nature, and of itself void of all Defilement, and being a ripe Body and mere Tincture, needs no other thing, than this, that its inward Parts be turned outwards, and its outward Parts inward, by that Philosophical introversion, and that so its external Yellowness may betake itself inwards, and the internal Redness may come forth outwards, or (to speak the more clearly) that the manifest Yellowness may be hidden, and the hidden Redness may be manifested. But now, the Case is not thus with this our black Earth, and impure SATURNINE Mineral; in the which, both good and evil, pure and impure, Poison and Medicine lye mixt together: Insomuch, that it is altogether necessary to separate the black and gross impurity, from that noble and tender Medicinal Pearl, and to reduce it unto Fixity.

A. With gaping Mouth, and open Eyes, do I even swallow down your Discourse, and yet I cannot pierce into the Foundation of the whole business, nor understand it. You speak of such an artificial Concentration, and which is beyond all my apprehensions, for I never in all my whole Life time heard anything of it, as far as I remember, much less read ought concerning it.

B. I will set before you then a Similitude,
that so you may the readier understand the knack,
Examine well, and consider, Wine, Ale, or the Lees
of Wine; and by your searching you shall find, that
there is in them but very little of the
Corroborating Spirit, or Soul, the reminder is
nothing else but mere impurities, destitute of all
Virtues. Now if some Physician or other should
administer to his Patient a large Cupful of Lees, to
cheer and comfort his heart withal, would you not
accuse him of ignorance, and Folly? For though there
is something of comforting Virtues in the troubled
Lees, yet it is impeded, or clogged, by the great
quantity of the dreggy Lees, for rightly performing
its proper Office. But put Case, that it could
display its own Virtues, yet nevertheless, even the
admixed impurity would also exercise its own
accustomed Evil, and so the Good would be confounded
with the Evil, or rather be quite over powered by
the same. For this reason, there is nothing more
necessary, than that the good be separated from the
evil, before anything be used about making MEDICINE.
Therefore even as there are Workmen to be found,
who, by the vulgar Distillation and Rectification,
do separate the Heart-refreshing Spirit out of the
sordid Lees of Wine and Ale, and bring it unto use,
give the remaining filthy dregs and useless Faces to
the Hogs to eat: Even so do we (the Lovers of the

175

SPAGYRIC Art) separate the pure Virtues and Powers of things from the gross, sordid, and noxious impurities, afore we administer them to our Patients.

A. As concerning these Sayings of yours, I am clearly of your Opinion, and withal do stick close to that common Proverb, which sayeth, Speak that which is truth, eat that which is baked, and drink that which is clear, if you be desirous of a long Life:

For my part, I delight to have the Kernels, and purged or cleaned Fruits, and willingly leave the husks and shells to the Swine, that are delighted with Bran and Chaff, and their own Dung.

B. I am extremely satisfied, that I have already (thanks be to God) seen the whole Work, and have a sufficient through understanding of the whole Basis of the same, viz, that the purer and subtler part is to be separated from the more impure and grosser part, and, by an often repeated Separation, and artificial Concentration, to be advanced into a most subtle, most pure, and most noble Quintessence, if anyone has a mind to produce effects of some great moment, or to do more than common things. And now, as you have learned, from the words of our precedent Conference, the way of making an universal Medicine out of common Gold: So have you likewise again received, from this our present discoursing together, the Confection of the universal Medicament, out of our SATURNINE MAGNESIA, which is the ROOT of the vulgar Gold. But, although both of these Medicaments are, as PER SE and singly alone, very excellent and effectual enough, yet notwithstanding it plainly, and clearly appears unto our kin, that the power and Virtues of both the universal Medicines admit of being promoted unto a far higher Degree; the which thing we see is possible to be done by the means of Conjunction, and is to be thus understood. When a Seed is implanted into its own growing Root, it attracts out of such a Root, which is of its own kind, or Nature a far better, and more nourishing Alimentary juice, than out of the gross Earth. For example, the Seed of a RAPE ROOT being put in the Earth, produces RAPES of the same bigness that the RAPES were of, from whence the Seed came: And as often as you shall commit that Seed (by sowing) to the Earth, you shall, notwithstanding, never have greater and better RAPES than those, from whence the Seed was taken. But now, if the Seed of some RAPE be sown into some other RAPE which is in the Earth, and which hath grown unto the half of its bigness already; that Seed will sprout forth, and grow up like as the other Seeds planted in the Earth will do, but yet sooner, because it finds a better Nutriment in its own Mother, than that does, which is planted in the bare

Earth: And hereupon, it must necessarily produce a greater RAPE ROOT, and from the greater Root will proceed a greater Seed. And this is the reason, that there are such great grown RAPES, which make every Body wonder at them: Some such I have seen, that weighed ten, twelve, yea and twenty Pound weight. If you have a mind to try, sow a few RAPE Seeds, throwing them into the Earth, which when they shall have grown to the bigness of an Egg, take a wooden sharp stick, and make therewith a small hole, even into the middle of the said RAPE, and then put there-unto your RAPE Seed, and stop up the hole with soft (clammy) Potters clay, that so the Rain may not get in and rot the Root. Thus now, that Seed will also (as being in its own Root) grow up out of this RAPE, existing and growing in the Earth, and will dilate that its Root, and advance it to a greater bigness. The Reason's this, because it can attract unto itself a better and more convenient Nourishment from a Root of its own kind, than out of the bare Earth. And if you shall practice thus with other Seeds likewise, you will effect the same, as with this.

N. B. From the Seeds of small RADISHES, implanted into great RAPE ROOTS, there grow huge RADISHES. This kind of Propagation may not unfairly be likened to a Mother's suckling her Infant, which attracts and draws its Nourishment from the Mothers

Breasts: But the Mother itself receives her
Nourishment from the Fruits of the Earth, and that,
being changed into a sweet Milk, supplies the
Infant, whereby it is sustained and nourished. But
that I may lay the whole business afore your Eyes,
by a more full Declaration, I will yet farther add
one Similitude more. Take you some wild and not yet
ripe Stock of a PEAR, APPLE, CHERRY, or some other
Fruit-bearing Trees, such, I say, as is not yet full
grown, but is as yet constituted in its first tender
Age, take off from it all its Boughs, which bear
(but) a wild and sower Fruit, and Cut it all off
even unto the bottom, and to the Stock standing out
of the Earth plant a SCION, plucked off from another
Tree that does not bear wild Fruit, into this Stock,
betwixt the Bark and the Wood of the same, where
'tis cut off with the Saw, and fence it well with
Wax, or tenacious Clay, & etc. against the Rain: If
now you shall thus do, and do your work Gardner-
like, that fruitful SCION will draw unto itself the
wild Juice, out of that wild Tree or Stock, and make
it better, so that it will no more bear any wild
Fruit, but such as that Tree did bear, from which
the SCION was plucked. If now, so small a SCION of
some cultivated and fruitful Tree can so change the
wild Juice of a wild Stock, that it becomes far
better, and more noble: Why, should not Gold do the
same in our Metallic Tree, when engrafted in Lead as

179

in its proper Root? Verily in my Opinion, there is not a better Earth (from which that golden Seed may attract unto itself an Alimentary Nourishment, and so multiply itself even to infinity) to be had, than its own proper Root. But yet, with this Proviso, that all the gross and degenerate Boughs be first cut off from that same gross Root, for so, it will the easier and readier change its wild and degenerate Juice, (when joined to the golden Seed or CION) and so will be changed into a far nobler, and produce Fruits a thousand fold. You see, my Friend! What good Will I bear to you, in that I have so faithfully opened unto you all those things which I thought necessary to be known, for the perfecting of so great a Work. Ponder well in your mind, upon the Similitudes of the RAPE and SCION, and believe that what I have here spoken is not casually and at Random. For they have more hidden under them, than they seem to show for, and then you would imagine. Length of time will open your Eyes, which are as yet blinded, as to abundance of things, (if God permit).

A. Surely, Sir, I cannot choose but wonder anew, when I accurately consider the things you have declared unto me, of changing the wild Nature, of the wild Vegetative Fruits into a tame, mild, Property: And that such a Transmutation of the ignoble Nature, into a more noble One, takes place also in Metallic Affairs. They are verily such things, as are of weighty Concernment, and most worthy a most accurate Consideration. But, indeed, who can sufficiently enough consider of all the Divine Miracles? Blessed be his most holy Name for evermore.

B. AMEN.

A. I give unto you, for your courteous instruction, most hearty thanks, and remain obliged unto you all my whole Life. And thus with what hath been said, I commit both you and myself to Gods protection.

B. My dearest Friend, I have now performed the Promise I made unto you, and am not meanly delighted, in that you have understood the meaning of those things, which I have said unto you: But yet I cannot but admire, that you are not Covetously desirous of knowing yet more, and that you do not Crave an yet more prolix unfolding of more matters. For you well know, that you cannot every day have the Enjoyment of my Company.

A. For those things, which I at present received from you, do I return you most hearty thanks, another time God will vouchsafe more: The greatest desire I have at present is to set about so great a Work, and to have the Fruitation of the hoped for Fruit? If you are so minded, and bent upon doing Friendly Favors, I do request you, that you would oblige my Brother by your good turns, if he shall hap to come unto you, and Petition for ought at your hands, for you may assist him in some small, yet good, ARCANUM: For he hath been stupid, or inapprehensive enough hitherto, and much needs some accurate instruction. But what shall I say? The sick looks after the Physician, he that is thoroughly well has no need of him. Such things as are hard to be understood exceed his Capacities: The more easy things are more commodious. And so again, Farewell.

The Corollary.

I have taught in this Dialogue, that the WHITE LAC VIRGINIS (after its being extracted, by Distillation, out of the Black MAGNESIA, and after its Exaltation, in Virtue and Efficacy, by Rectification and Concentration) IS TO BE FREED FROM ITS SUPERFLUOUS HUMIDITY, AND YET THE DRY MATTER IS TO BE MADE PERMANENT IN THE FIRE, BY THE GRADUAL OPERATION OF THE FIRE. Now for the more accurate Declaration sake; these things which follow shall be yet farther subjoined.

That Fixation, if it be to be perfected by the (bare) help of the common Fire, requires a long and tedious time, so that there's no reason for a Man to persuade himself, that he can finish the same in one year's space. I speak by my own experience, for I myself have tried, and find that it cannot be, that one year should suffice for the finishing of this Fixation, for indeed it requires a much longer time. For after that, I had so far advanced the Matter, that it had passed through all the Colors, as to flow when put upon a Red hot Plate, to insinuate itself there into, like Oils penetrating into a dry Hide, yet was it not fixt enough nor constant, nor served it for the tinging of the Metals, but when a vehement Fire was applied thereunto, away it went in fume; but yet not without an evident Demonstration of the possibility of the same. For as much therefore, as it does not yield that satisfactory Fruit, (and Success) and seeing that such great Labors are not undertaken, without the expectation of some Profit, and that the hoped for Fruits cannot however be gathered afore that the said matter is promoted to a perfect Maturity, and consequently dreads not any the most vehement Storms of the Fire anymore; any one may easily conjecture, that there needs (as I said afore) a more tedious space of time for this Fixation, if it be to be done and perfected by the common Fire of Coals. But now, he that has the knowledge of the secret Fire, of the Ancient Philosophers, such an one will much easier, and speedier, arrive unto the wished end of the Operation. The Nature of the vulgar and fugacious Minerals doth very difficultly and slowly admit of that Fixation, which is made with the Fire of Coals:

And this I was unwilling to conceal from the diligent Searcher after the Secrets of Nature; yet farther adjoining this Admonition, (viz.) that a very profitable Medicament may nevertheless be prepared in a shorter space of time, and an appearance made of the admirable, and highly delightful Variation of abundance of most delicate Colors. For the first Color that appears is like the black head of a CROW, presenting itself to view like the Color of black Glass. This blackness going off by little and little, gives place to the White, and ponderous Mass; which is called by the Philosophers, the WHITE SWAN, and not without reason, because that selfsame Matter is not so compact and Stone like, as that black CROWS HEAD, but is porous, and not much unlike unto a kind of heap made of abundance of small white Feathers. When this Whiteness is turned into a Yellow, those Feathers vanish, and the Mass returns to its former Compactness, and resembles the form of a yellow Stone:

Of which if you put a little piece upon some red hot Silver, or Copper-plate, it will at first stand like a Red blood, and afterwards penetrate the Copper-plate, and tinge it both within and without with a white Color; but yet, somewhat brittle as yet, and yields in the CINERITIUM, or CUPEL, some Silver, and operates in Medicine somewhat effectually, like the white Stone, but yet weaker.

It likewise pierces into a Silver plate, like as Oil into a Skin, and tinges it with a yellow Color, which being separated by the CUPEL, and dissolved in AQUA FORTIS, leaves excellent Gold in the bottom. I have not as yet made any farther Progress on the Operation, being quite tired, and weary of spending any longer, and more tedious time thereabouts, which however was necessarily requisite to the perfecting of this Tincture with the Fire of Coals. But yet I have by me all the Colors as they follow on after each other, which I can show unto any one; whereby they may see with their Eyes the most evident possibility of Nature: To which end also I am minded to preserve those Tinctures by me, that they may be an everlasting Memorial of so great a thing, unto my Posterity: But for my part, I will commend (to every one) that shorter way of bringing the work to the wished end, by the Mediation of the secret Fire of the Philosophers: Concerning which, the following Dialogue, and the little BOOK OF FIRES, treats. For the immature First ENS of Gold, cannot be excocted (or digested) into the desired Tincture by anything more easily, than in its own proper secret Fire; and not by a Coal Fire.

And that it may clearly appear, that I have written the Truth, I will send unto some of my Friends (God willing) some of those white and yellow Stones, to be used not only in Medicine, but in

ALCHEMY too; that so they may make trial and
experimentally find, that Tinctures have a Power of
bettering, and amending Metals, afore they have
arrived unto the half part of their Fixation.

The Third Dialogue

The Third Dialogue, or Conference, betwixt B. and C., treating of the true Universal and particular Medicine of the ancient Philosophers, (extracted) out of such Gold as is yet fugacious, or Volatile, and immature; (and is) to be ripened by their secret Fire; which Operation is by them styled, the work of Women, and play of Children.

C. Good morrow heartily, my Friend! I wish you a good and prosperous Day.

B. I wish you the same, whence come you to us so early?

C. I have very earnestly for these several days waited for this hour, that I might see you, and have the Fruitation of the most sweet Fruits of your Courteous instructions. My Brother A. sent me hither, and told me, that you had given him leave to send me unto you this day. I do therefore Friendly request you, that you would put a good Construction upon this my Boldness.

B. (My Friend!) I do not in the least take it ill, your Brother told me, that you were as yet ignorant and unskillful, in the more subtle ARCANUMS and Secrets; and that you therefore needed not ought else, but some pretty easy Secret, which might be easily understood, and performed with small

Expenses, and yet be profitable and beneficial unto you.

C. Indeed, Sir, to tell you the plain Truth, such ARCANUMS as are profound ones, and to be penetrated into by a subtle Meditation, and which are of great moment, do far outreach my duller apprehension: And therefore I do not at this time desire anything, more than this, viz, that I may obtain something that is not costly, and yet may effect so much in Medicine and ALCHEMY, as may serve me to live a little more commodiously and plentifully, as to Food and Raiment. And that you would be pleased, to render me a Master of this my desire, is my humble request unto you, again and again.

B. You do very prudently measure out your requests, according to your own Capacity. And indeed, sometimes, those things which are not Costly, nor are difficult in their Operation, are more profitable to a Man, than those things are, which they would fain get by the expense of a great deal of Charge, of a long time, and hard Labors: I will grant you your Request, and therefore hearken.

C. (Sir, I thank you), and do listen.

B. Have you never read, or else heard from others, that those most ancient Philosophers, tell us that their universal Work, is not only most easy to be done, but withal no ways Chargeable. For they do openly Confess, that to the perfecting of their Work, a Man needs be at no more Expenses, than two FLORINS, and that the Labor itself, even from the beginning to the end, is nothing else but mere Woman's work, and Boys play.

C. You are pleased to offer me delicate Dainties, easy to be prepared, for I have not so much Money as is to be laid out upon those kind of chargeable and costly Labors: Neither will my Family Affairs admit thereof, viz, to spend my days in such sumptuous and chargeable Cookeries, and which are so full of most great Labors, and whose event too is oftentimes very uncertain. For I have heard some say, that the smallest Error may destroy the whole Work, and quite spoil it, and so grievously effect such as operate about such great Secrets, with a very great loss of long time, and great Expenses. And therefore such a Work, as I can attend upon without letting my taking Care of Family, and which will need the disbursement of but a few Expenses, will please me better than those other Costly ones would.

B. I readily believe that this very thing you desire will not only be exceedingly pleasing unto yourself, but also unto many others besides. The Expenses are but very small, and the Labors thereupon but little, so that each days Fire, which the Matter is to be set upon, may be taken Care of, in half an hours space. But now, though the Philosophers have made mention of very small Expenses, and have comprised them in the compass of a couple of FLORINS, yet is that saying to be otherwise understood. Those Ancients made use of the greatest FLORINS, viz. the RHENISH ones, and also the HUNGARIAN Crowns, each of which is of the value of five of our FLORINS. And if you thus understand it after this compute, I can easily show unto you the Truth of their Assertion.

C. I do even think as you say: Sure we must not expect any (such) things for nothing: I am content, (and therefore pray) let us proceed.

B. If our Work be called the labor of Women, and Boys play, it is expedient, that it be like unto Woman's work, and Boys play: For else the Philosophers would have used an unfit similitude. You well know, what Labors your Wife is chiefly busied about, and what her daily Labor is she employs herself in.

C. Yes, verily I daily see that she doth boil Food necessary for the Use of the whole Family, and being boiled set it upon the Table to be eaten. This Labor she performs, at least twice every day; when Dinner and Supper is ended, she doth wash the Dishes, Pots, Goblets, and other Vessels, and cleans them, and makes them fit to put other new Food in, and to be served out to the Table. Besides, too, this is her Office and Care, if haply a Pot be broken, or cracked, whereby it is made unfit to hold moisture any more, to substitute (in defect of Iron Pots) a new Pot made of Potters Clay, in the stead thereof: Such and the like Labors, as those, are in our Country called the Woman's Work.

B. Well, be it so: I will likewise show unto thee, such a Labor in CHEMICAL Operations, as resembles this. Therefore, like as the Female Sex do first wash the FLESH, RAPES, FISH, POT HERBS, ROOTS, APPLES, PEARS, or other things with pure Water, which they mean to boil, and then put them into the Pot, and pour thereunto as much Water as is requisite, and place it over the Fire, and boil it so long, till all the crudity, or rawness being vanished, the Meats become grateful to the Palate, and pleasant, and easily digestible by the Stomach. (So do we). They do likewise sometimes pour Wine upon Flesh and Fish, instead of Water, and add as much Salt as is convenient, together with some Spices, or odoriferous Herbs, by which they give the Flesh and Fish a most excellent Taste. But yet we must not forget Salt, above all the other Spices, or Seasonings, and odoriferous Herbs, for it Corrects and maturates the Flesh, Fish, and other hard Meats, more than other Spices. For we can well enough want these other Spices. For we can well enough want these if they are not at hand, but as for Salt, there is always need of that, about the boiling of Flesh, Fish, and other Food. If therefore Flesh, or Fish, are to be boiled well, then Salt water is requisite; and as for all the other Additions of Herbs, and odoriferous Spices, they only serve to give it a good pleasant Taste, and make it acceptable to the Palate, and to the Smelling. For the Flesh and Fish when boiled or stewed, do by their Magnetic Virtue attract so much Salt and grateful Savor, and Virtue, as they need: And that which remains, stays in the Water. Now the curious Dames do shut the tops of their Pots very well with their Covers, lest the efficacious Vapors should be forced away in fume by the boiling, and not stay with the Flesh or other Meats. But the careless Housewifes do not much regard the covering of their Pots, from whence it comes to pass, that they lose good and sweet Spirits, and then they fill up their Pots with new Water, by which doings, the Flesh, or Fish, do not get so sweet a Savor, as they would have, if that efficacious Water had been kept in and conserved. Some Women that are yet more curious, and diligent about their Cookery, do put upon their Pots, wherein they boil their Meats, such a Cover as hath a fold in it, by which the Collected sweet and odoriferous Vapors may distil down into an under—put Vessel, which being thus gotten, they keep by them, to refresh and cherish with them, such as are weak and sick, when need requires. Others, to free themselves from this kind of Labor, do add as much Water as need is, together with Salt and Spices, to their Flesh, and so boil it by little and little, shutting in the Vapors with a Cover, which else would go away, and the Meat taste of burning. And by doing thus, they are not necessitated to pour on any new Water, though this slow boiling takes up more time than that, which is done by a strong and incessant Ebullition: I would have you well to observe these things, for 'tis not without cause that I utter them. And now let us examine the other similitude, and see what those Boys Plays are, that so we may afterwards accommodate even them too, to our Philosophical work. What therefore do you see concerning the Boys Plays, with what things, and after what manner do they Play?

C. How can I tell? They play as their Please to let them, or as they can get opportunity of Playing: As for myself, I do not grant my Children so much Liberty, to play when, and how they let themselves. I send them to the School and to the Church, and sometimes I allow them one hours Play for Recreations sake, nor do I allow them any other Play but at Bowls, (or Knickers) by which they moderately stir their Bodies, and exercise themselves, and Concoct their Meat, and this is far more profitable for them, than if they were constrained to sit always at home, without any exercise at all: Other play than this, I allow them not. Cards and Dice are unfit Plays for Boys, they are many times very hurtful to those of riper years, especially when by the too much abuse of them, they do so unprofitably waste their precious time, and cannot tell how to use a mean. I have indeed otherwise seen Boys, that meeting with some Sand get (there out of) bright Stones, and play with them, but yet this is not usual.

However, there is no play more frequent amongst Children, than that of Bowls, (or Knickers) which play they daily use, whatsoever time they can steal, to that purpose. Yea both at their going to School, and returning from School, you may see them presently busied about their Rubbers, & Knickers-play. They are very hardly restrained therefrom. If they want Money to buy the Bowls, or Knickers, they get a little piece of Potters Clay and moisten it with Water, and make up their little Bowls, or Pellets in their hands, and harden them in the Fire; which I remember, when I was a Boy, I often did. And besides this Boys play, no other is known to me.

B. Very good, you have hit the nail on the head: And now let us see whether or not the ancient Philosophers have (after the manner of Boys) played with small Bowls, or Knickers? And whether, or no, they have boiled their work in Pots, with as easy a

Labor, as Women do. For of necessity they must have hit on doing after this wise, else could they not have compared their work to the Labor of Women, and play of Children. So then, if we are to imitate Woman and Boys in our Operation, what Matters are we to make use of, for our boiling, in the stead of Flesh, Fish, and other Meats; and what Water is it, that is to be poured thereupon. For if we are minded to do any good effect, 'tis expedient, that we likewise know, what those Matters are, which admit of being boiled into a Maturity in our fiery Water, and these verily must be such, (seeing they are to be MATURATED BY BOILING) as have a great affinity with the said Water: Forasmuch therefore, as our Water is of a METALLIC Nature, and yet all the Metals do in a manner arise, or proceed (in the Earth) therefrom, and are even at this very day advanced, by the very same (by the help of the Terrestrial and Central fire) by little and little unto perfection: All that we have to do is, to imitate the simplicity of Nature, which will never seduce us, for so without question, those most ancient Philosophers did do, who having borrowed their wonderful Work from Nature herself, do advise us to do no more, but to follow Nature, and to begin there, where Nature left off, and to ascend higher and to make that perfect, which is as yet imperfect. God hath prefixed unto Nature her bounds, which she cannot transgress (or go beyond). But Art, doth much excel Nature, and performs those things which Nature cannot accomplish: Yea more, that which she can hardly do in the Earth in a thousand years' time, Art effects in one year, and this is easily confirmed by many Testimonies. Now as to the Generation and Maturation of the Metals, Nature uses a most simple or plain way, a very slow one, but yet safe. From thence arises the Error of many a Man, who do not follow Nature, but the guidance of their own fantastic Brains, never effecting ought of good, but remains always Novelists in the same, what Labor soever they undertake, and what Expenses soever they are at: Although the ancient Philosophers do by their many Admonitions set afore us, that most simple Course of Nature for us to imitate; and they have especially hinted to us, that their Work is so simple, that should they but openly and clearly have treated of the same, even the Women would deride it and say, that the Male kind had learned their Art from them. Yea, it is so very vile a Work, that no Body would be able to believe it, and upon this Account, the Philosophers have done their utmost, to hide and obscure the Art the most they could, least they should be condemned by the proud deriders, (who Soaring aloft seek after things too high for them) and be accounted for Cheats and false Writers. And this is the main and chief Reason, why this Art being so wrapped up in darkness of a most profound Silence, lies hitherto hid from the whole Troop of Sophisters, and such deriding Fellows. SENDIVOGIUS (as we have already several times hinted) doth expressly say, that he had oftentimes declared the whole Art, to not a few word for word, unto whom, that Art did nevertheless seem so very vile and mean, that they could not at all believe any likelihood of Truth, in his most true words, and so left the Work unattempted. The same SENDIVOGIUS doth also say; That had the most skillful HERMES, the most quick witted GEBER, and most illuminated LULLY been again alive, and beheld our Laboratories stored with so many, and such various Instruments of Glass, Earth, Iron, and other Matters, and such several Furnaces, they would be ravished into a most high Admiration, like so many Boys, and would be but as it were our Scholars, as concerning these Vessels and Furnaces, all which however, we have learned from their Writings, but yet we are destitute of that most excellent Work which was wrought by them in so simple a Way, and it hitherto flies our subtitle and acute Wits. And my dear Friend he also tells us, that we should fly aloft into the lofty Air with our Wings, for the Work is simple, vile and abject, the which, you may sooner comprehend (or feel) with your hands, than apprehend by the subtlety of your Wit or Cogitations.

C. All those things may very sufficiently serve to rid us out of so great a Labyrinth; but I pray, Sir, how comes it about, that we do sotishly persuade ourselves, that those things are so very difficult, which notwithstanding are so very simple, vile, and abject.

B. It is indeed, to him that has knowledge of the same, an easy, vile, and simple thing: But very difficult and intricate to him, that laying aside the way of Nature, thinks himself able to learn so great an Art out of Books, which (by their leave) though, seems a thing almost impossible to be done. For the Philosophers have so prolixly, intricately, and obscurely described the whole Mystery, that their so prolix and dark Writings would sooner lead a Man from the true and right way, so far off are they from reducing him there into.

C. I myself find, that this is most certainly true, for I never heard as yet of any Man, or read of any, that learned the Art out of Books: But that almost all of them who were skilled in the same, do Confess, that they became Masters of the same, either by Divine Inspiration or Revelation, or by the help of some Friend. There's nobody can contradict those things which you have here induced, for the Confirmation of your Opinion. And now, Sir, let us set upon the Work itself, and diligently pray unto God and wait for his Blessing.

B. Content, hearken therefore attentively.

C. So do I.

B. Did you never find in the reading of the Philosophers, that all the imperfect Metals may in a PARTICULAR way be promoted to the perfect Maturity of Gold or Silver, by their dry Water which wets not the hands: But being not content with this effect, they have promoted the first ENS of Metals (by the help of their occult, fiery, and ripening Water) to a more than perfect Constancy, and Fixity in the Fire, and have concentrated it to the form of Gold.

C. I remember, that I have read of such things as these, though by reason of my unskillfulness, and ignorance, I could not understand the least Particle of their meaning: For I am altogether ignorant of that kind of wonderful, and yet, to every Body well known Water: And so shall still remain until it be shown (and pointed at) with the Fingers.

B. Look here, here's a piece of admirable Water which is everywhere in all places easy to be found, yea, in the poorest Country-men's Houses, nor doth it deny any Man, the Possession and having it.

C. Who would have ever believed, or thought, that there was any good hidden in so vile and abject a Body?

B. (I wonder) wherefore (it is) that no Body can think of this Matter, seeing it is evident, that in our Water the first Entities of all the Metals are abundantly hidden, as in their own proper Seed, out of which they are generated in the Earth, and ripened into perfection. Those first Entities are but extracted out of this our Water, by (peculiar) Magnets. And now, like as every Metal hath its own Magnet, even so every one (of them) hath its proper Magnet, by which it is concentrated into a narrow Room. I will declare unto you the Truth of this business, by an example taken from the Metals. If you have a Water fully impregnated with Copper, and you desire to have Copper out of the said Water, you will easily bring this to pass, if you shall but put in to that Water, which holds in it the Copper, a piece of clean Iron, the which Iron, (as being the true Magnet of the Copper) will attract unto itself, all the invisible and palpable Metal. If Silver be dissolved in any Water and made invisible, and you would again have it, put into the Solution a Plate of pure (or clean) Copper, which (after a Magnetic manner) will gather together (or draw out) all the Silver in his own (Silvery) Body, and make it visible and palpable. Now when Gold is dissolved in some Water, and largely dispersed, (as I may say) and you would again have it (in the form of Gold) then put in some MERCURY or ARGENT VIVE in that Water, and boil it a little therewithal, (as was done above, with the Copper and Silver) and you shall presently see all the Gold to be attracted, and gathered together by the MERCURY, insomuch that there will not remain ought of the Gold in the Water, because it follows the Attraction of its own Magnet, MERCURY. These METALLIC and MAGNETIC Operations, are a sufficiently manifest Information unto us, and do point out unto us (as with a Finger) the way, of the extracting, not only good Metals in a PARTICULAR manner, but also far better things than Metals, (viz, the Tincture, or form of Gold) out of our stinking Water, in an UNIVERSAL way, by such Magnets, as are fit and apt in Nature for this Extraction. Another similitude we have from the Earth, and Rain water, with which the Fruits of the Earth are moistened:

Put you in such an Earth moistened with that Water, as many Seeds as you please; each of them, will (by its MAGNETIC Virtue) attract unto itself, its own like, for its Multiplication, and will leave to the other Seeds, to attract each of them its like also. If now, by this similitude, the Scope or end I aim at, may be made manifest unto you, there is yet hopes you may be helped: If not, I do not see, by what means you can be succored, forasmuch as it would be too tedious here to use many other Circumstances. For when we are certainly assured, that the first ENS, or the very Form of Gold is plentifully hidden in our Water, we do by very good right seek after that best part, viz, the form of

Gold, and leave the other first Entities of the rest of the Metals, in the Water. And now I will show thee another similitude: Dissolve in one and the same Water, SOL. VENUS, MARS, JUPITER, MERCURY, that so you may, have all these Metals commixt together in the same: If now you desire to extract the best of them, viz, the Gold there out of, what hurt will the other bring you, if they remain in the Water. Such therefore as the Magnet is, which you put into that same Water, such is the Metal too, that you shall extract. If therefore we particularly seek SOL and LUNA in our Water, it will be expedient, that we put unto this spiritual SOL and LUNA their proper Magnet, unto which Magnets, they (being precipitated) do stick on, and are by little and little fixed (on thereunto). And now if we seek after something better than SOL and LUNA, viz, the form and Tincture of Gold, it will be necessary, that we also apply such a (suitable) Magnet, which may draw out nothing else but the Tincture or Form of Gold, which being precipitated (there-out of) may be fixed. And thus have I here told you all those things that are necessary to be known. If you are minded to extract in a PARTICULAR way, SOL and LUNA out of the Universal Mineral water, you must then put unto them their Magnet, viz. an AMALGAM of Copper, and ARGENT VIVE: For the Copper draws to itself the spiritual Silver, and the MERCURY the spiritual Gold, out of the said Water, and brings it unto a Fixation (or Corporeity) with itself, (or, as itself is). But if you are minded, or desirous, of getting some better thing, than the Gold itself, or Silver, is, viz, a Tincture, you must then needs adjoin thereto its own peculiar Magnet, for, without it, you cannot effect anything at all: But that you may know, what the true Magnet of the Tincture is, I say unto you, THAT IS THE COMPEER, OR COMPANION OF OUR WATER, AND NOT OF THE METALS:

For like seeks its like, as the Philosophers say, Nature rejoices with Nature, Nature overcomes Nature, Nature retains Nature:

More than these things I have told you, it is not needful for you to know: Consider therefore very accurately what I have said, and beg of God by your Prayers a blessing, which if you do, you shall not err, but yet you will not be all of a sudden Master of what you desire. All these things have their determinate times, like as a Grain of Wheat sown in the Earth, which requires a time to ripen in, nor doth it wax ripe afore the time appointed for Maturation be come about. Follow you the Advice left by GEBER, and do not hasten on your Work, by any the least hasty speed: For he tells us expressly, that all hastening in our Work is of the Devil. And as concerning what is to be known, and what a one the Student of so noble an Art ought to be, you will find described in the Fifth Part of my SPAGYRICAL PHARMACOPEIA. And what think you now? Can you thoroughly understand me?

C. Yes, Verily, I do well enough understand those things which I have hitherto heard from you; But yet this Woman's work, which you began a Declaration of, is not sufficiently clear unto me, nor is that Boys-play, which is done with small Bowls or Knickers, I do not thoroughly understand that neither, viz, how it may be compared with the work of the Philosophers. Were but these things made clear and evident unto me, I would Rank myself amongst the number of the Masters of the Art.

B. Well then, come let us go on and see, by what means the Philosophers are wont to make their little Bowls. (Note!) Like as the Boys make use of Earth and Water, for the making of their Pellets or Knickers: So likewise will we use our Earth, and our Water to the making of our small Bowls or Pellets: Ours, I say, not the vulgar and common Earth and Water, for they are unprofitable as to our Work. But tis indifferent, and all one, whether we take Yellow, Red, or white Earth, because all of them are of one and the same Nature. According to the Color of the Earth which we use, will the Color of the small Balls we make, be: We have here at hand a threefold Earth, a yellow Earth of Gold, a white of Silver and a Red of Copper. This threefold Earth

204

will we moisten with our Water or MERCURY, and so make up a Paste or Mass of the two, which the CHEMISTS call an AMALGAMA. This Paste will we wash with pure Water, grinding it so long 'twixt our Fingers, till there be no appearance of any farther impurity, and that it admits of being easily washed, or broken with the Fingers. Being thus well washed, we will put it in a Skin, or Cotton, and tie it firmly in; out of which, we will squeeze forth all the MERCURY with our hands, and separate it from the said Earth, just in such a manner as ARGENT VIVE is separated in AMALGAMATIONS, or METALLIC Masses. The MERCURY being separated, we will take out our thick AMALGAMA, and make small Bowls, or Knickers thereof, and of almost the same bigness, that the Boys Knickers are of. These Bowls we will expose to the Air, for about some twenty four hours, and by this time they will be grown so hard, as to resemble Stones, in hardness. And now, being thus made ready for boiling, we will put them into our Water to seethe: But yet in the seething, there must be an accurate regard had to some skillful handling the Matter, if you would have it perform its Operation without Error. This Art will I faithfully open unto thee, lest you err, and so bring damage upon yourself: These Pellets, or little Bowls is thus made of our Mass, are not to be thrown into the Pot filled with Water, afore the said Water boils: Which

205

boiling, you must throw them in one after another, and they will presently harden themselves, and cover themselves over with an hard Crustiness or Skin; by which they will be prevented from sticking to one another, and from coming altogether into one Mass. For if you first of all put your little Balls into the Pot, and then afterwards pour cold Water upon them, and so set them on the Fire that they may be boiled into their Maturity you will spoil your whole Work. For before the Water in the Pot be hot, all the Pellets being dissolved, would run together in one Mass, and so would not admit the Humefying, or moistening of their inside Parts, and so you would turn all your Pellets by your seething into a mere Pouse or Pottage, whereas they should remain whole in all their Parts, as well internal, as external, for fear of drawing the Water in. But that you may have an accurate Account of all these things which I have said unto you, I hope you will set about the Work, (or thus) I hope you will do your best, heedfully to regard all these things which I have said unto you, and I suppose, that you have sufficiently understood the things which I have already spoken: But if so be, you do not yet apprehend the meaning of these things, I will Counsel you, by what means you may learn this Operation at home from your Wife. When you are come home, bid your Wife to make you some little Balls, or Dumplings with Flower and Veal. And heedfully observe, what Course she takes, about making such Balls, that so you may by the same way learn to deal with your METALLIC Pellets. First of all, you shall see that she puts some pure Flower into a deep Dish or Platter, and having put it in, she works it into a Paste, or Mass with Cream, or the purest Milk. Then she admixed some green Herbs finely minced, and some Spice melded among, and sprinkling some Salt thereupon, she mixes them with the Paste made as aforesaid, to give it the sweeter Smell, and Taste. Of this Paste, she will afterwards make little Balls, of what bigness she pleases, which Balls she does not put in the Water afore it boils. When therefore the Water seethes, she throws in, one Ball after another, each of which, as soon as ever it feels the fervent heat of the boiling Water, will presently cover itself over with a Skin, whereby they will be kept from sticking one to another, and running into one Mass, and returning into such a Paste as they were in, before their being made up into Pellets: Whereas, now each of them may be kept in its own Form and be encompassed all round with the Water, and be advanced unto a Maturity, or readiness, by a due seething:

And now, when you have seen this Operation of our Wives Cookery, I do not question, but that you will be well enough skilled in this Cooking Art.

C. My dearest Friend, I do friendly request you not to take it amiss, in that I cannot obtain any longer from downright laughing, when I hear that our work hath such a corresponding likeness, with the Art of Cookery: Your so faithful Instructions have already abundantly satisfied me: I have very often seen my Wife busied about the Cooking of such Balls, and myself have also delightfully fed upon those kind of Dumplings, made both of Veal, Eggs, and Spices, and also of Flower, Milk, and green Herbs. But I fear me, that when my Wife shall see me making those Balls or Pellets, and boiling them in a little Pot, she will laugh at me, and say, that I learned my Skill from her.

B. 'Tis no Matter, you have no reason to regard either the tattling of your Wife, or of all Men whatever: For they know not what they do, it is enough for you, that yourself know what end it is, you do anything for; Think you, that if other Wiselings and Know Littles should see you working with such little Balls, they would not deride and mock you. But don't you at all mind their unprofitable Prating, leave the shrill-sounding Geese to their own loud chatting, and follow you my Doctrine, and wrap up this our Cookery Art in the darkness of Silence: Which if you do, you need not fear of being mocked, or laughed at by either Women, or Womanish—Men.

C. I have now (praised be God) learned enough: But yet, there is one thing I am ignorant of, and that is this, by what sign I may come to know, when my Pellets are well enough boiled, and what Fire they are to be boiled in. The Fire of Wood and Coals, I know is used by the feminine Sex for to boil withal, but whether or not, the same be necessary and conducive to our Operations, I request you to inform me.

B. Have you never seen, what proof Women have to know, when their Dumplings are well boiled? They are wont to take one out, and cut it in twain, that they may see, whether or not the inside, as well as the outside of the same be so well seethed, as that the Flower is not any more tenacious or Clammy: Do you even the same, and sometimes put a piece of one of your Pellets you take out, in the CINERITIUM or CUPEL, and that will show unto you, how much Increase of SOL and LUNA hath particularly added its self to your Balls, in that time of the boiling, and how long they are, as yet, to be boiled. Now you know, that all these things are to be searched into, by your own Meditation and Trial, because it cannot possibly be, that all things can be so very clearly set afore one's Eyes, as to need no farther meditating thereupon, and inquisition there into. After this manner may you boil in one Pot, with one and the same Water, two, three, or more little

Bowls, of different kinds, as some made of Flower, green Herbs, Spices, Flesh, Eggs, Fish, and other things, and so, after the seething of them, you make, take forth one kind after another, and PARTICULARLY apply them to your Use. For these things are to be understood concerning PARTICULARS. But if you have a mind desirous after the UNIVERSAL Medicine, then must you enter upon a certain way of almost a whole entire year, which is necessarily requisite to the finishing of its Operation. For our Magnet, whose Task it is to extract the FORM of the Gold out of our Waters, doth as yet groan under its immaturity, and therefore needs no small space of time, for the extracting the TINCTURE out of our fugacious and combustible Waters, and fixing it with itself.

C. These Words of yours, by which you mention so long a time, do not a little frighten me. Our Wives can boil their Dumplings enough in one hour's space; what will such a continued boiling cost? I would be glad to redeem it, (or, to be excused) for the price of two Golden pieces of Money (or Duckets).

B. I should tell you, that you are of the Off-spring of unbelieving THOMAS, for you heap upon yourself, by your needless incredulity, such heaps and Loads of Cares. Don't you remember, that I told you at first, that the Charges of the whole work;

from the beginning to the very end, do not exceed two Golden pieces of Money, which they call Duckets. And that I do not at all tell you an untruth, I will expound it unto you by an evident Demonstration. When you shall heat your well covered Pot, that so your Water may not vanish away in fume, with the small fire of a Lamp, how much I pray will such a Labor cost you? Put Case, that some Pounds of Water cost you some ASSES or STRIVERS, and the Magnet doth also cost you some STRIVERS, (ten ASSES are a ROMAN Penny, which is Seven pence half a Penny of our Money) and now how many Pounds of Oil will there need to nourish that so little a Fire? And although you should spend forty, fifty, or more Pounds of Oil, may not you well say that you shall finish the whole work, for the charge of two Golden Duckets. Well! Are you yet Content?

C. You do now again somewhat encourage my mind, which had almost fainted, by telling me, that the Matters necessary for our Work are sold at so mean a Price. But there's one thing still that doth not a little trouble me, and that is, that so much time is required to the Fixation. All the other things are as well as I could wish: But, I would fain have had that shorter work of three hours, or seven days.

B. O thou Dreamer, what have I to do with thee? What? Doth that space of time, wherein such excellent and most profitable Fruits are to be hoped

211

for by thee, seem too long? What dost thou think to get without length of time? Good things are not wont to offer themselves without Process of time, as the common Proverb tells us. Meanwhile you may follow your Vocation, nor needs there any other Labor, than that you look to your Lamp Morning and Evening, and see how the Fire is. And I pray, are not the Countrymen constrained to wait their time, wherein to reap, and again to receive the Fruits which they committed to the Earth? And though they have sown their Seeds afore the approach of the Winter, yet they cannot reap them again from the Earth, sooner than the next following AUGUST, which then rewards and recompense all their hard Labors. But now the waiting so long time does not tire them; for they patiently expect the time of Harvest. Thus likewise are you to do, but if you are greedily desirous of sooner making ready your Pellets, or Balls, by the boiling, you may I Confess, have a sooner ending of your expectation, and that on this wise, viz, by a stronger Fire, which may make your Water boil without any ceasing, but yet in this same way of Operating there doth again happen this trouble, from that strong and incessant seething, viz, that your Water being without any intermission resolved into Fume and Vapor, is always lessened, and you must of necessity be always pouring in more new. Use which of these two ways you please, for you are clean importunate and troublesome enough unto me. I will not, for the time to come, take on me to instruct any more such Disciples as you are. What do you think, that if that short work of three days, or of seven hours were known unto me, I would presently reveal it unto you? No! But yet I am not gotten to so high a Degree of knowledge, as to profess myself a Master of so great an Art. I do believe though, that such things are possible to be done, but I deny, that I myself am able to do such notable things. And now, go home in God's Name, and diligently and seriously meditate upon all these things: You have heard enough, and my time will no longer permit me to confer with you. If perhaps, one or two Errors should put a stop to your proceedings, you may again come unto me, and ask me thereabouts. Meanwhile I commend you unto God, and pray him to bestow his Blessing upon yourself, and your Labors.

C. Now am I contented, nor know what farther I should ask:

I am sorry, that I have so much troubled you by my dull Brains, and beetle Head, and been so importunate: Nor know I, how to requite your deserts; God will reward you with Life eternal, I shall go home full of joy, and bear a glad Message, and Tidings unto my Family. And I pray God to be at all times present with us, to our Help and Succor.

B. AMEN.

The Corollary

I have, in this Third Dialogue, made mention of a certain secret fiery Water, which can. ripen the Volatile and immature Minerals, and Metals: And herein I have principally regarded a PARTICULAR Transmutation. But forasmuch as a PARTICULAR Melioration of the Metals requires as much time, and no fewer Expenses, than the UNIVERSAL itself does, I would here commend to the Sons of Art the UNIVERSAL work, which is to be preferred before a PARTICULAR one. Such things as we have perhaps omitted in this Dialogue, the Description of my fourth secret Fire will sufficiently supply the defect; to which, I refer the friendly Reader. He will there see and understand, that the greatest part of the whole work, and the very hinge of the said Operation, consists chiefly in the true Vessel, in which our Matter is to be ripened; and without the knowledge of which, there can never be anything done to purpose, Which Vessel, seeing that all the Philosophers have covered over, and hidden with so great a Care and Diligence, and have involved it with such obscure Clouds of darkness, I should do amiss, yea, most extremely amiss, should I lay it open, and bring it from out of those dark inwraprnents, into the Light. Thus much only I say, that it ought to be such a Furnace, and the Vessel itself such a one too, as in which, all the Chemical Operations, viz. Solution, Putrefaction, Distillation, Sublimation Cohobation, Ascension, Descension, Circulation, Inceration, and Fixation may be Perfectly shown unto an HERMETIC Disciple, or Learner, in one hours' time, in one little Furnace, in one Glass, and in one Fire, all which must not cost more than the value of one quarter part of a Dollar, and is all done without any changing of the Glasses, or putting to, of the hands of the Operator, These are indeed such things as exceed all the belief of the whole World, but yet they do not exceed GLAUBER'S faith, nor suffices it him only to believe, but he can also effect the same, that so other incredulous people may believe likewise.

N. B. On a certain time, as I was familiarly discoursing with a learned Man, concerning such great and incredible things, he presented me with these following Verses, aptly agreeing to this Matter in hand.

Thy Surname (Glauber!) given was, as unto FAITH relating
Yet by good right it should have squared unto thy Operating.
Faith's Objects are invisible, but yet, such things, do you,
As would at first be scarce believed, produce by Art to view.

FINIS.

215

NOVUM LUMEN CHYMICUM

NOVUM LUMEN CHYMICUM:

OR, A

NEW CHYMICAL LIGHT.

Being a Revelation of a certain new invented secret, never before manifested to the World.

Whereby a clear and inextinguishable light is set before the Eyes of the blind World, and, as it were, palpably demonstrated, that good Gold may be found and attained with profit, everywhere throughout the whole World, as well in cold as in hot Regions, so that in all those places, where Sand and Stones are found, a Man cannot set his Foot, where, not only Gold, but also the true matter of the Stone of Philosophers is afforded.

Reader,

I will now address myself to the discovering of the wonders of God, not hitherto heard of; to wit, that throughout the whole Earth, no less in cold than in hot Countries, where there are Sand and

Stones to be had, good Gold may be extracted from thence with profit. Because a Man cannot set his foot in any place where Gold doth not exist. Nevertheless Chalk, or Limestones, are to be thence excepted; because they only seldom or never contain anything of Gold, otherwise all sandy and rocky Stones, all Flints of what color whatsoever, as also all Sand, fine and course, all sandy Stones upon high Mountains, in deep Valleys, in the bowels of the Earth, in Ponds and Rivers, and lastly, all the Sand on the Sea-shore, none excepted, although it hath as yet obtained no color, but be white and clear. Which indeed will seem incredible to very many, but is asserted by me as a sincere truth, which I never found by reading or hearsay, but have proved by many and various experiments. Some of which I will here set down for the confirming and demonstrating the truth of my Writings, that by those, all Men of a sound mind may be able to discern, that my Assertion is no Dream, nor fabulous trifle, but sufficiently founded upon experience in the light of nature.

Therefore I would not have you judge of those things which thou understand not, nor cannot at present apprehend, but remember that the wonders of God triumph with magnificence and power. But search out those things in every part before you presume to interpose thy immature judgment. And although in the proof you should be mistaken as may easily happen to one making trial, yet do not impute the error to my Writings, but rather to thy own unready wit. For I will here write nothing but what I have oftentimes performed, and can yet perform at any time. Consult other searchers after this matter, among which I think all cannot err, in a matter so easy, that even a Boy of ten years of Age may apprehend it, for what belongs to the possibility of it, but it will be evident to many of them with me, THAT THERE IS GOLD IS ALL SAND AND STONES, THROUGH ALL PLACES OF THE WORLD. But there is no need that I should here show the way of extracting it from them, in a great quantity or large use, but rather I shall beware, that I put not so sharp a Dart into the hands of my Adversaries, to wound myself, for I have published this to gratify candid Friends, not Counterfeits, but least of all the Compilers of notorious Libels, in which opinion, I will remain and acquiesce.

We will now proceed to the Trial whether those things which I have here written be true, and are able to abide the Examine of the Cupel, to wit, that in all Sand, there is good Gold.

The first Specimen of Probation.

RECIPE one Ounce of white Sand or Flint, which you esteem to be altogether void of Gold, with which mix three parts of MINIUM, or of any other Ashes of

SATURN. Put this mixture into a Wind Furnace, or to the Bellows; let it flow well together for an hour, and it will turn to a yellow glass; suffer it not to stand too long lest the glass of SATURN perforate the Crucible; and run out into the Ashes. Pour out the glass and powder it, and mix it with half its weight of SAL ALKALI or Pot ashes, put this mixture into an Iron Crucible, into which you have first put some bits of Iron, or old Nails; give fire and cause the matter to flow, and the glass of SATURN will be continually reduced by the MARS, and at length return into Lead: Pour it out into a Cone, and the REGULUS of SATURN will sink to the bottom; and the Sand or Flint will be uppermost as a dross. After all is cold, take out the REGULUS OF SATURN, which hath drawn to itself so great a roughness and blackness from the Iron, that it cannot easily be cupellated; which you may remedy thus; put this black REGULUS OF SATURN into a crucible in a Wind Furnace, cause it to flow well; and if there be of that one Ounce, cast upon it a Drachm or somewhat more of Salt—peter, and make them flow together, then the Salt-peter will attract the roughness from the SATURN and turn it into SCORIA. Which being poured out and the REGULUS OF SATURN separated from the SCORIA, it will be white and tractable, and easily runs upon the Cupel. This I call washing. But if you know not how to perform this Lotion, which yet in itself consists in an easy labor; put thy black and rough REGULUS OP SATURN into such an Instrument (as they call TREIB SCHERBE) or a close Cupel; under a covering (or Muffle) and suffer it so to be desiccated for the space of half or at least a quarter of an hour, pour it out and separate the SCORIA from the REGULUS; which will be white and tractable. But the Lotion by Salt-peter is much the better.

Of this REGULUS OF SATURN, and of that same Lead of which the Ashes of SATURN were made, take equally the same weight, exactly weighed with the lesser weights, put each of them apart upon a dephlegmed Cupel, suffer them to run; compare one with the other, and you shall find, that SATURN which was blown off with the Sand or Flint to leave behind it a grain of Gold, but on the contrary, the common SATURN will only leave a grain of Silver. Who now can deny, but that the grain of Gold proceeds from the white Sand or Flint? For no Gold could come from the Salts. This Specimen of Probation is a palpable Argument, that Gold is contained in all Sand and Flints of what color soever. But that there is no difference between one Sand or Flint, and another, so that there is no more Gold in one than another, I do not assert, for (in that respect) they greatly differ, Also that the Sand of hot Regions contains more Gold than that of cold Countries, is in no wise to be doubted, as every prudent Man will easily think. For we have set down this Specimen of Probation, only to the end that you may find Gold to be in all the Sand and Flints of the whole World.

The second Specimen of Probation.

RECIPE one part of white Flint or Sand, with which mix three or four times as much Salt of Tartar, or of any other Alkali, which mixture put into a Crucible, so that it be not above a third part full, because this mixture in the melting might rise up and run over the Crucible; let it stand half an hour, that it may be well melted, and it will be turned into a white transparent glass. Pour it out and dissolve it in water, or rather in a LIXIVIUM, and the Sand or Flint will be dissolved, and converted into a thick Liquor.

In this Liquor digest for the space of an hour or two, half an Ounce of Filings or Raspings or rather thin shavings of Lead, and the Lead will extract the spiritual Gold from the Water of Flints, and will thence grow yellow: Which being taken out of the Water dry, cupel it, and you shall find a grain of Gold. Also cupel another half-Ounce of the same Lead, and you shall find a grain of Silver, as is commonly wont to be had from all Lead, from which you may be certain that white Flints and Sand, contain spiritual Gold, which becomes corporeal with Metals.

A Third Specimen of Probation.

Dissolve SATURN in AQUA FORTIS, and pour into the solution Salt Water, and all the Lead will fall from the Water like a white Powder; mix three parts of this edulcorated and dried CALX OF SATURN with one part of Sand, to which add half so much of the Salt of a LIXIVIUM or other Alkali; which mixture put into an Iron Crucible, into which you have first put some bits of Iron, which being covered, suffer it to melt for the space of fun half an hour, until the sharp Spirits of Salt which were in the Lead be mortified by the Iron, for then the SATURN is reduced and rendered corporeal, which being poured out into a Cone, the REGULUS of SATURN falls to the bottom, which you may wash with Salt-peter as we have taught above, or purge it from the Faeces on the TREIB SCHERBT (or Muffle) weigh it exactly by the lesser Probatory Weights, weigh also as much of any other common Lead, which is not melted with Sand, cupel each by itself, and that Lead which was melted with the Sand, will leave a grain of Gold in the Cupel, but the other common Lead leaves only a grain of Silver. Which sufficiently proves that there is Gold in all Sand, none excepted. But I do not assert that all Sand contains so much Gold as that it may be thence extracted with profit. Because some Sand is Poor, other rich, another richer. Therefore whosoever intends any profit by this ought beforehand to know the difference of Sand and Stones, that he may not lose his labor. The which may easily be found out by experience, to wit, if you do nothing in great Quantities, before you have made trial in lesser, that is, such as may be performed by the lesser Specimens of Probation.

A Legitimate and Infallible way and manner of finding out and proving every Flint, Stone, and Sand, whether it contains much or little Gold.

RECIPE four ounces of Sand, Flint, or other Stones or Rocks, which you have a mind to prove; heat them red hot in a Crucible, and cast them so into cold Water, where they will become tractable, and may easily be brought to fine Powder, to those four ounces of powdered Flints, Sand or Stones, put into a small Glass Cucurbit, pour two ounces of AQUA REGIS, that the Sand may be well moistened; set the Cucurbit upon warm Sand for half an hour, and the AQUA REGIS will attract to itself the Gold which is contained in the Sand or Flint. To which if two ounces of hot Water be poured into the Cucurbit, and mixed by a strong shaking it together, and filtered through a brown Paper, the Water with the Gold goes through the filter, and the Sand will remain in the Paper, to which if you pour a little common Water,

and let it run through it will take out the residue
of the Gold, which yet adheres to the Sand, which is
to be added to the first. Upon this Solution if you
pour a LIXIVIUM, but rather Spirit of Urine; that
LIXIVIUM or Spirit of Urine, mortifies the AQUA
REGIS and precipitates the Gold which is in it in
form of a yellow Powder, from which the Water is to
be poured off, and the SOL to be edulcorated with
fresh Water, and dried, and that very warily,
because Gold of this sort, when it grows hot,
fulminates so that it is shattered into minute
parts. To this fulminating Gold if you add a little
common Sulphur, and make it red hot in a Crucible,
it doth not fulminate. If you reduce such a CALX of
Gold mixed with Borax in a clean Crucible, you will
find how much Gold that four Ounces of Sand or Flint
contained. N.B. Except the Sand or Flint should
happen also to be impregnated with Iron, which would
render the SOL brittle and pale. For in that case,
the SOL precipitated from the AQUA REGIA, and burned
with Sulphur would not flow with Borax, because the
Iron being mixed with it, would adulterate the
Probation: But if you cupel such Gold partaking of
Iron, with SATURN, the Iron may be separated from
the SOL, and the proof will void of error. N.B. That
Sand and Flints may indeed be proved another way,
but seeing this which we have here prescribed, is
easy to be practiced, we will rest in that.

But this one thing is also necessary to be known, viz. It is indeed true, that in all Sand and Flints there is Gold, as the above mentioned proofs do witness; nevertheless know that there is a difference between native, corporeal, and solid Gold, and the volatile spiritual first Ens of Gold. For the corporeal Gold is easily extracted by the help of corrosive Waters or Salts; but the spiritual no so. And because corporeal Gold is always elicited by the benefit of the above mentioned Probation it may thence happen, that although in white Sand there should plainly be no corporeal Gold, nevertheless by the help of the aforesaid Probations it would be thence elicited, indeed not much, nor no more than what the Lead which was used for the proof, did contain of Silver: because the LUNE in the melting or dissolving hath attracted to itself the spiritual Gold, from the Flint or Sand, so that it is thence tincted and transmuted into Gold. Which is manifest in that the grain of SOL doth not exceed the magnitude of the grain of Silver, which is gotten from other common Lead. But if besides the spiritual, there be also corporeal Gold in the Sand, the grain of SOL will necessarily exceed that of LUNA, for the Silver which was contained in the SATURN doth not vanish into Air but remains. And because it doth not remain the same Silver, but passes into Gold, that change is ascribed to the

226

notable Operation of the first Ens of Gold, or spiritual Gold in the Sand. Wherefore whosoever intends to extract Gold from Sand, Flint or Stones with profit, ought to know certainly before by smaller experiments, that besides spiritual, they also contain corporeal Gold. For I have written this Book only upon the account of extracting from them corporeal Gold, leaving the spiritual Gold to Philosophers that of it they may make their Stone. But necessity required that I should lightly touch at those few things. For if according to my prescribed Probations, any one in working should have found a grain of Gold of equal quantity with the grain of Silver, he might wonder in his mind, which way the Silver had made its escape, seeing that he had found Gold instead of Silver. Wherefore I thought it necessary to show the reason of that, for the taking away all scruple from every ignorant mind. Wherefore whosoever seeks Gold from Sand or Stones, let him choose such, from which corporeal Gold may be extracted with profit. For they will afford him such profitable fruit, as the white Sand denies. But that for the performing my Specimens of Probation, I have taught to take white Sand, I have done it for this reason, that I might make it evident to everyone that there is Gold in all Sand, but that it cannot be extracted from all with gain. For white Sand or Flint is very often void of corporeal Gold, but never of spiritual, by which Silver may be converted into good Gold. I might have passed by this admonition with silence, seeing that those things are not necessary to be known by the vulgar, in as much as corporal Gold will abundantly satisfy their desire; but a Philosopher neither wants, nor seeks corporal Gold, but only spiritual, to prepare of it a Tincture. Add, that to him it is sufficiently known, in what subjects the first Ens of Gold is plentifully contained. Hence, although the first Ens of Gold is also in white Sand, as is manifest by the foregoing proofs, he uses not that, but rather seeks such Stones for his work, as contain much Tincture. Neither also doth any true Philosopher so tie himself to this or that subject only, that he will not elicit his Tincture from any other, inasmuch as it is evident to him that the first Ens of Gold may be found in all things in the whole World. For wheresoever there is Sulphur, there may also be had the first Ens of Gold, whence a Tincture may be prepared. But it is evident that in all Vegetables, Animals and Minerals there is Sulphur. For the matter of the Stone of Philosophers, everywhere offers itself, so that the poor as well as the rich may attain it without cost, according to the saying of the Philosopher; THAT THEIR MATTER IS EVERY WHERE OBVIOUS, AND PREPARED

WITHOUT MONEY, TROD UNDER FOOT AND THROWN TO DUNGHILLS.

So the true Philosophers speak and write. But Sophisters, who falsely assume the name of Philosophers, Wandering up and down from Court to Court, where they offer their service for the acquiring the Philosophers Stone, by their lying Fables, persuade credulous Noble men, that the matter of the Stone is to be sought in this or that place or mountain, as in HUNGARY, TRANSYLVANIA, the Forrest of HERCYNIA, THURINGIA or BOHEMIA, or in the Rocks of SWEDEN, or NORWAY, and elsewhere. Others again persuade credulous Persons, that the Stone of Philosophers, can be made of nothing but common Gold, and therefore require of their Patrons not only some ounces, but some pounds of Gold, for their own greater profit. One of those Pseudo-philosophers was lately here at AMSTERDAM; who by persuading certain Merchants, tricked them out of two and twenty Marks of Gold, thence to make the Philosophers Stone. He indeed put the Gold into the Vessel according to his own manner, before those Merchants, giving to it a continual Fire. But after much time being elapsed, when they expected to reap much Fruit, he, who had played the Philosopher, privately withdrew himself into I know not what Country, and would not return to take out the two and twenty marks of gold. Therefore the Merchants themselves being about to take them out, found those solar Birds were flown, and had left only the empty nest. Which empty nest as yet lies in a place in this City, nor doth anyone know how to convert it to his own use. But if the Merchant had given to that impostor in hand, only two, or at the most three ounces of gold, they had not received so great a loss. But as they unadvisedly credited his fine Clothes, and his finer words, so they were intolerably deceived.

A true Philosopher wants not so much gold for his Medicine, inasmuch as if he brings but one half ounce to perfection, it may suffice for his whole life, and may be in his own power to do the same again as often as necessity shall require, so that he will have no need to run up and down from one to another who are greedy of gold, to blemish the noble art of Chemistry, and bring it into hatred with the whole World.

Some years since, when I had written of the Prosperity of GERMANY, and from a good intention had taught HOW WINE AND CORN MIGHT BE CONCENTRATED, AND IN CASE OF NECESSITY, MIGHT BE COMMODIOUSLY CARRIED FROM ONE COUNTRY TO ANOTHER, AND THERE SAFELY KEPT IN GARRISONS OR FORTRESSES FOR FUTURE USE; fearing that at some time while the Shepherd slept, that most ravenous Wolf the TURK might break into the GERMAN Fold, and cut the throats of what Sheep he could not carry away. Against which, FARNER with his accomplices, hath set forth notorious Libels, and hath everywhere traduced me for a false Prophet: Whose Calumnies I have not opposed, because he hath sheltered himself under the protection of Princes. Moreover, this FARNER, with a consort of ruiners of fame, or good name, have reproachfully wounded my Honor, by falsely accusing my Writings, and traducing them for Lies: And seeing that almost all of them abide in great Men's Courts, and have not put their names to their invective Papers, I have not indeed been able to reach them, But yet at length those Men of darkness, who have concealed their names, that they might give the greater reputation to their slanderous Papers, do come to be more and more known to me. Besides at last the event hath showed those things to be true, which I had predicted should happen; to wit, that the depraved Christians should be chastised by the TURKS and TARTARS. But it is to be pitied that also Men of better note; should be ensnared and captivated by the common judgment, by reason of that Company of the Devils attendants. For the unerring truth declares, that the righteous scourge of God, will not be taken away from us, before such spoils of a Man's good name be restrained by the Magistrate.

Now, to return to our purpose, to demonstrate, that not only Gold, but also somewhat more rare; to wit, a true Tincture is hidden in Stones, and by the help of Art may be extracted from them, which the Ancients have hinted at in these words: AURO QUOD MELTUS: JASPER, etc. And that there is much Tincture contained in the Jasper, I have long since inculcated in other places of my Writings. PARACELSUS highly commends RED TALC, GRANITES, ANTIMONY, and LAPIS LAZULI, adding that Tinctures or the first Ens of Gold may be gotten from them by the help of sublimation. Moreover the first Ens of Gold, may also be found in other Stones, especially in the HEMATITIS, SCYTHIS, MAGNESIA OF PEDMONT, SMIRIS, and others of that kind, in which it is so fixed, that there is nothing wanting to obtain it but only the way of extracting, and giving it ingress by SOL. On the contrary, the first Ens of Gold is indeed in Vegetable, Animal, and Mineral Sulphur, Marcasites, and Antimony, and that abundantly, but so volatile that Stones are to be preferred to them.

And although my intention in this Book extends no farther than the extracting of Gold out of Stones, Sand, and Flints, yet nothing hinders but that in a few words, I may also show, that in Stones, there is not only fixed Gold, but also volatile, whence a true Tincture may be perfected. Which Tincture I have not as yet made, but nevertheless I am certainly persuaded that it is no wise impossible to be done. For whatsoever can render the first Ens of Gold, in Stones, volatile, and draw it out by Distillation, he will continually obtain a gradatory water, whereby he may coagulate every running Mercury into good Gold. But he that knows how to join this volatile Gold with corporal SOL, to unite them and procure ingress, may thence expect a far greater good. For the first Ens of Gold is more conducible for the preparing Tinctures, than corporal Gold, as one of the Philosophers hath signified in the following words: THOU CANST NOT MAKE GOLD OR SILVER, EXCEPT THOU SHALT OBTAIN THEIR FIRST ENS. The first Ens of Gold, which is hidden in all Vegetables and Animals; coagulates Mercury also into a yellowness, but not constant; but if it shall be fixed, it also coagulates and fixes with constancy, but not before. Therefore it is most certainly true, that where Sulphur is, there is also the first Ens of Gold, and where there is the first Ens of Gold, there is also Tincture. Therefore seeing that Sulphur may be found, in all things in the World; even in the smallest Herb, it follows that also from every Herb, Wood; Stone, or Bone, a true Tincture may be prepared. Whosoever will believe; let him it matters little to me whether it be believed or not, I think my self-satisfied in that I have not put that light which I have received under a Bushel, but have exhibited it to the World for information. For the light conduces nothing to him who is blind, and resolves to remain so. You shall find more in my third Century, as also in the sixth part of my PHARMACOPEIA SPAGYRICA.

How it may be known whether Sand, Flints, and the like Stones, being pregnant with Gold; contain much Gold, or but little.

White Sand and Flint, contain the least Gold of all; which Sand indeed is never without Gold, although it cannot be extracted from it with profit. In the white and red there is more Gold than in the white, nevertheless not always so much as will pay the cost. In the yellow, brown, or black, there is commonly much Gold contained, whensoever veins of white or yellow Sand run through them, and especially when such Sand shines with abundance of small golden sparks, closely adhering to each other. In like manner that Sand is rich in SOL, which is like to Talc, or when Stones are found in it, in which there is red or brown Talc, even as SOL is contained almost in every Talc, nevertheless in some more; in other less.

All River Pebbles(which although outwardly they appear white, yet when heated red hot and broken, are stained with yellowness; and contain SOL) are in extracting sufficiently rich in Gold, Green, yellow, or blue Stones being transparent like horn (called in the GERMAN Tongue HORNSTEIN) are also for the most part fruitful in Gold. Also all red; dark colored and black Flints contain Gold, with which Iron is frequently admixed, which in the extraction is drawn out with the SOL and weakens the MENSTRUUM, and renders it unprofitable.

All Quarries or rocky Stones, whether they are found in the Earth after the manner of Ducts or whether they lie in the open Air, when they are colored, contain Gold. Every HAEMATITE, and that which is a kin to it, the SMIRIS, the Granite, and LAPIS LAZULI; all contain Gold, nevertheless always one more than another; so that some of them are rich in Gold.

All Granites contain Gold; and besides Gold; also the first Ens of gold; but they are endowed with so hard a body; that AQUA FORTIS cannot exercise its power upon them. But there may be a remedy found, by whose help their extraction may be instituted.

In every golden Stone called CHRYSOCOLLA, SAPPHIRE, RUBY, AMETHYST, and HYACINTH, there is gold, but difficult to be extracted, All Flours or things fluxing, which are applied to the Mines of SOL and LUNA, for the bringing their Ores to a flux or melting, whether they be of a violet color, or purple, yellow; red; or green, are all endowed with unripe and volatile gold, Which if you shall make red hot, the color evaporates like a green, yellow, or red smoke, leaving the Stones white. But if any

Man shall know how to intercept, and detain such fugacious SOL, he may therewith coagulate MERCURY into gold. In like manner from all Stones, in which is the first Ens of gold; a green water may be extracted by the help of Distillation; in which MERCURY coagulates itself into gold. The Ancient Philosophers have called such a green water, their GREEN LYON, which devours gold, and of it prepares a Tincture for LUNE and MERCURY.

Concerning this matter, I could indeed say something more, did not the avarice and improbity of Men who seek nothing but the damage of their Neighbor, restrain me. Hence all those in which any light of God shall arise, ought diligently to beware that they communicate nothing of it to wicked Men, although they put on the shape of an Angel. For Faith is now nowhere kept among Men, as these following words of a golden Alphabet do witness.

In God alone repose thy trust,
With Men's presence be not beguiled;
God only keeps Faith, is Just,
Which from the World is quite exiled,

If many years ago, I had known those things which I now do, it would have profited me much in being aware of the dissembled sanctity of impostors. But what is past cannot be recalled: Let these things at this time suffice to be spoken of the knowledge of Sand and Stones, every Man may look further for himself, and search out and learn more. If any Man finds any good from what I have written in this Book, let him give God thanks, and be mindful of the Poor; if not, let him look upon himself as unworthy to be partaker of it. For I have here written more perspicuously, than any Philosopher hath done before me, in which I rest, It now remains that we should say something concerning the easy way, and incomparable invention, by whose benefit SOL may be extracted from Sand and Stones, in great quantity and with a considerable gain. Which invention hath been hitherto unknown to the War Id, and it is almost impossible that a better should be found, But that the searcher into Art may see that this new invention of mine, is of all the most easy for the extracting of Gold from Sand or Flints, and that the World hath never known the like, I have thought fit to show in what this COMPENDIUM consists, and it is thus. It is sufficiently known with what labor and costs the ancient Workers in Metals, and even those at this day, have extracted, and do extract Gold from Sand and Stones, viz, when first they burn the Ores or Stones, then grind it in peculiar Mills accommodated to that purpose, with ARGENT-VIVE and water, where the ARGENT-VIVE draws to itself the SOL, and the

Stone goes away in the washing, and the greatest part of the ARGENT-VIVE is strained from the gold through Leather, which may again serve for the same use. Then in Iron Retorts they drive out the residue of the ARGENT-VIVE, which yet adhered to the SOL, which is saved; then they melt the SOL. In which operation they lose much of the ARGENT-VIVE, which in the grinding and washing goes away; so that often times the charge of the MERCURY is no less than the gain of the gold. Therefore by this way nothing can be gotten from a poor Mine. But this is the easiest way they have to separate SOL from its Ore or Stones.

Another way is, when they mix the washed Ore (which the GERMANS call SCHLICHT) with its weight of Litharge, and melt it with Bellows, then cupel the REGULUS OF SATURN, and so they obtain the SOL which was in the Ore. Which way of melting is also dear, because much Lead is lost: But yet they are obliged to follow this way, who know no better. Now I will compare my way with this, that it may be evident which of them is the most easy and profitable. As for my method of extracting, it chiefly depends upon four singular COMPENDIUMS, by which the operation is rendered easy. The first of which is a water of small cost, which may be copiously prepared, without Distillation.

The second is a singular Metal, of which the Kettles or Pans are made, in which the Stones or Sand is boiled with the water of small cost, and yet are not corroded nor consumed, And when the water shall have dissolved and imbibed the SOL in the Sand, the Ore, Sand, or Stones, with the water are to be taken out of the boiling Vessel, and put into another Vessel adapted to this use, with a bottom all over perforated with small holes (like a Colander) and first covered with inside, with a thin Matt, then the water will drop down through the Sand and Matt into another Vessel placed under it; moreover more hot water is to be poured upon the same Sand, and let to run through it, which water will also extract the gold which as yet remained in the Sand. And after this Elixiviation, the Sand (which is now of no use) is to be taken out of the filtering Vessel, and this labor is to be continued so long, until no more Sand is to be extracted.

The third COMPENDIUM is this, when the LIXIVIUM of Gold, is collected to a just quantity; a singular Water of small cost, is poured into the Solution of Gold, whence all the SOL is precipitated from the Solvent, the Solvent by inclination is decanted from the CALX OF SOL, and may again be used for the like extraction, seeing that it still retains its strength, being nothing weakened by the precipitation. Therefore we may use this Solvent a long time. And that which is lost by pouring too and again, is of small charge, because it may be recruited again by the like cheap Water. But if one should mortify that dissolvent with other contrary LIXIVIUMS, to precipitate the SOL, as otherwise is wont to be done, and I have taught above, about the proving of Sand, what a loss should we undergo, if at every time we should destroy our Solvent? Besides such an extraction is costly and laborious, when made in glass or earthen Cucurbits.

But this extraction is almost of no charge, seeing that it may be perfected in great Kettles, and the Gold thence precipitated without loss of the Water, This extraction of Gold is like the extraction of Salt-peter from Earth, where the Workmen also elixiviate one part of the Earth, by the help of Water, which Earth they then throw away, and in room of that put other Earth into the Vessel, which they also elixiviate or wash, and that so often till they have elixiviated all their Earth impregnated with Salt-peter. And by the same reason we elicit our Gold from Sand, as they make their Salt-peter.

The fourth COMPENDIUM is this, that the precipitated CALX of SOL, after the Water is filtered from it is taken out of the filter and dried, and by the benefit of a certain singular good flux, not at all costly, is reduced, so that in the melting nothing of the Gold is lost.

In those four COMPENDIUMS the whole work of our extraction consists, as a building standing upon four Pillars, one of which failing, the whole structure is ruined. He that knows those four COMPENDIUMS may boldly enter upon the work, to extract SOL from Sand and Stones; for then it will be a work of profit to him, otherwise not, which I have declared for the information of every man.

But that I have made none of those four COMPENDIUMS manifest, let no man wonder, because I have been sufficiently hurt by the slanderous Forgeries of cavelling Detractors. For when by reason of their dull apprehension, they could not understand nor perform those things which I had written openly and plain enough, they falsely traduced my Writings, as abounding with lies.

Let these things suffice to be written at this time, concerning the extraction of Gold out of Stones, Sand or Flints; which I doubt not but will be of use hereafter to many indigent Persons. For although all men should apply themselves to that extraction, yet they would in no wise incommode one another, seeing there are Sand and Stones everywhere obvious to all. And also the Salts for extracting are so plentifully afforded, that nothing is wanting but a man to put his hand to the Work. But someone may here object and say: I do indeed believe GLAUBER, that in the Sand and Stones of the East and West INDIES, AFRICA, SPAIN, ITALY, FRANCE, and other hot Countries, much good Gold may be found, but who will remove his dwelling thither to extract it? There is no need that any should go dwell in AFRICA, or either INDIES, thence to fetch us Gold, seeing that it may be had sufficiently in EUROPE, and may be found in all cold places. It is well known, that in many parts of GERMANY, Gold is elicited from the Sand as well of small, as great Rivers and Lakes. But that more Gold may not be had in GERMANY, which is temperate, than in cold NORWAY, or SWEDEN, and less in GERMANY than in FRANCE or SPAIN, I will not deny. Therefore we need no Commerce with thirsty AFRICA, or AMERICA, as that with great peril of body and mind, we should seek Gold thence, seeing that we have it in every Country of EUROPE, and that not only underground in the Veins of the Earth, whence it is to be dug out with very great trouble, cost and labor; but it is also everywhere manifest upon the Earth, where it is much more largely and easily acquired. The most famous Monarch of Philosophers, PARACELSUS, in his Book of the Vexations of ALCHEMISTS say: THAT MORE GOLD AND SILVER MAY BE FOUND ABOVE THE EARTH, THAN IN ITS PROFUNDITY, AND THAT OFTENTIMES A COUNTRYMAN THROWS A STONE AT A COW, WHICH IS OF GREATER VALUE THAN THE PRICE OF THE

COW. Which thing indeed is certainly true,
nevertheless whosoever will not with GLAUBER believe
it, let him remain in his unbelief. In ESDRAS, we
read these words: THERE IS MUCH EARTH OF WHICH
POTTERS MAKE THEIR VESSELS, BUT A SMALL QUANTITY OF
POWDER OR DUST OF WHICH GOLD IS MADE.

Which words are commonly thus interpreted,
that, by the words powder or dust of which Gold is
made, the Writer hath understood the Stone of
Philosophers. The which is very agreeable to truth,
but yet no Earth is found void of metals,
nevertheless all are not so rich, as to afford any
profit in the extracting, On the contrary all Stones
and Sand (although every of them doth not contain
corporal SOL so largely, as to be thence extracted
with gain) rejoice in the first Ens of Gold, or such
a CALX by whose benefit Gold may be made. Which
powder, if we know how to extract from them, we may
make it better than Gold itself. But seeing such an
Aurific CALX is very largely diffused in Sand or
Stones, and cannot be thence hammered out, but is
extracted by Art alone, therefore the blind Covetous
of Gold, with the Ignorant, will not believe it,
because they know not how to perfect it. Hence an
Art of this kind was by the Philosophers kept most
secret, where PARACELSUS speaking of the first Ens
of Gold, said that it may be elicited by
sublimation. BASILIUS writes that the preparation of

the universal Tincture may be compared to the distillation of a burning Spirit from the Lees of Wine, A comparison sufficiently perspicuous! For even as in a great quantity of Wine or Beer, there is hidden but a small quantity of good Spirit, and the residue is nothing but an unprofitable mud, but nevertheless that little Spirit, of which one spoonful is more to be esteemed than a whole pall full of Lees. By such means the Philosophers would have us extract by Art, the first Ens or form of Gold being very far dispersed in Sand and Stones, and concentrate or bring it into a small compass, of which, but as much as the magnitude of a Pease, is of greater worth than a huge Mountain of unprofitable Earth, Moreover, this I will not conceal, that throughout all GERMANY, in and about the Rivers may be found Stones, which are sufficiently rich in Gold and Silver, and moreover if you break them into pieces, you shall find within them little Holes and Caverns, which abound with a yellow or brown Powder, which if any one shall take out, and melt it together with Borax, he will acquire SOL mixed with Silver. But as yet, I never saw any man, who knew this sort of stones, and much less that golden CALX which is hidden in them. Which thing without doubt they have passed by, by reason of their negligence, not loving or seeking to know the physical MAGNALIA of God. I have found many the

like stones in the sandy Hills about URRECHT, and in
other Sand pits of this Belgick Nation, but more
about the Banks of the RHINE, and the Isle, as also
the Coast of the Southern Sea, some of which stones
I have yet by me, Also in the rough places of
WALAVIA you may find much Sand and Stones, which
contain corporal Gold. But there is no man who knows
anything of them. And this might be of great use to
Children, if they were placed for a time in the Shop
or Workhouse of some Artist, of whom they might
learn, whence in any case of necessity they might be
able to sustain themselves. But the rich, relying
upon their own fortunes, think that they have Wealth
enough for their Children, but if any adverse
fortune shall chance to befall them, as their Houses
to be burnt, or their Ships robbed by Pirates of
their rich Merchandizes, or be cast away by
Tempests, or their Detours break and run away in
their Debts, then they know not which way to turn
themselves; and because they have learned no Art,
whereby to get their living, they commonly
degenerate into men of a desperate life. For one
leaves his Wife and Children, and goes into the
INDIES, where not a few have become a Prey to wild
Beasts or Cannibals. Another, for a small stipend or
pay, sells his Freedom to fight by Sea and Land,
until like a mad Dog he is miserably slain. (I have
heard that it is a Custom in GERMANY, that a man may

play away, pawn, or sell his Liberty or Freedom, which being once gone, he becomes a Vassal or Slave to the public Service of the' Prince or State, during his Life). Others, (after they have consumed and wasted all their substance, and have learned nothing that is gainful in their youth, whence they might honestly maintain themselves and Families) betake themselves to a vicious kind of life, till they perish in it. To have truly learned Mechanic Arts, indeed helps much in fortunate times; but when the times happen to be troublesome and difficult, and many men in the same City exercise one and the same Art, one often hurts another, by eating the Bread out of his mouth, and reduces him to straits. But if a Physician knows somewhat besides the Profession of Physick, by which he may obtain a Living, he will have no need to make so many Visits to the Sick out of a pressing desire and expectation of getting money for his diligent attendance. Therefore Hermetic Philosophy and Medicine, with their Cousin GERMAN natural Alchemy, are the most excellent of all Arts, and will so remain to the End of the World.

Seeing therefore that such great Treasures (as we have heard) are hidden in the despised Earth, and in such subjects as are everywhere trampled underfoot;- wherefore should we not extract them, for an honest maintenance, and defense against the injury of the times? Why should we not leave INDIA to the INDIANS, and have regard to our own EUROPE, which abundantly affords whatsoever we need for the sustenance of Life? I cannot but say again and again, that were I but ten years younger than I am, I would not cease, but for the good of the public, would publicly teach, and demonstrate to the Eye, true Philosophy, Medicine, and Alchemy. But my glass being almost run, I must commit the care of this labor to others who are younger and have greater strength of Body.

Therefore in the meantime, whatsoever good I can do my Neighbor by wholesome Writings, I will not omit. I intend shortly (God favoring my design) to publish many secrets, hitherto unheard of. Nothing now remains but to close this little Treatise with,

THE END.

Glory be to God alone.

An Admonition to the Reader.

Whatsoever I have written in this little Book of the compendious extracting of Gold, out of Sand and Stones, is so true, that nothing at all is to be doubted concerning it: Nevertheless after this Tractate was gone to the Press, another way, and that much better came into my mind, by the benefit of which, Gold may be extracted with a much greater expedition, than by the help of the former. Because for the operation of this last way, there is no need of any Kettles or Pans made of a certain singular Metal, but the extraction may be made in quantity without any boiling, in such Vessels as are everywhere in hand, and may be had, so that one man in one day, may by an easy business perform the extraction of a thousand Pounds weight of Sand. Which method is indeed much to be preferred to the former. Wherefore I could not but also notify this way of extraction. What do you seek? If I shall find that those my profitable inventions are gratefully received, I will not neglect officiously to serve the public, in this present evil Age, and in the worse to come, by publishing the same. With which, benevolent Reader, I commit thee to the Divine Care and Protection, Dated at AMSTERDAM the 26th. Day of July 1664.

A Word from the Publisher

Thank you for purchasing this small work from The R.A.M.S. Library of Alchemy. During his lifetime, Hans Nintzel was dedicated to the identification, acquisition, study, retyping and, when necessary, translation of what he considered to be the most important known works on Alchemy. Hans was assisted by his sparse network of fellow Alchemists, all members of the Restorers of Alchemical Manuscripts Society (R.A.M.S.). I was an active member of R.A.M.S.

My goal is to publish all of the works originally made available through R.A.M.S. as photocopies. To facilitate this, I have chosen to have the books professionally printed. I also have a few titles that I intend to add to the original R.A.M.S. Library, selected by strict criteria established by Hans.

The works from the original R.A.M.S. Library are republished by R.A.M.S. Publishing Company in the collection, "The R.A.M.S. Library of Alchemy," with permission of the Estate of Hans W. Nintzel.

If you have a work on Alchemy that you believe should be a part of the R.A.M.S. Library, please contact me through R.A.M.S. Publishing Company.

Philip N. Wheeler

www.ingramcontent.com/pod-product-compliance
Lightning Source LLC
Chambersburg PA
CBHW082301200526
45168CB00017B/2307